The Gelfand Mathematical Seminars

Series Editor

I. M. Gelfand

The Gelfand Mathematical Seminars, 1996–1999

I. M. Gelfand
V. S. Retakh
Editors

Springer Science+Business Media, LLC

Israel M. Gelfand
Department of Mathematics
Rutgers University
Piscataway, NJ 08854-8019
USA

Vladimir S. Retakh
Department of Mathematics
Rutgers University
Piscataway, NJ 08854-8019
USA

Library of Congress Cataloging-in-Publication Data

The Gelfand Mathematical Seminars, 1996–1999

ISSN 1068-7122

AMS Subject Classifications: Primary 03C60, 05B35, 17Bxx, 17B67, 20B25, 22E30, 52B45
Secondary 03C13, 16Sxx, 20G10, 22E47, 22E99, 32G34, 33C80, 51M10, 52B10

Printed on acid-free paper.
© 2000 Springer Science+Business Media New York
Originally published by Birkhäuser Boston in 2000
Softcover reprint of the hardcover 1st edition 2000

SPIN 10645593

ISBN 978-1-4612-7102-4   ISBN 978-1-4612-1340-6 (eBook)
DOI 10.1007/978-1-4612-1340-6
Reformatted from authors' files in LATEX by TEXniques, Inc., Cambridge, MA.

9 8 7 6 5 4 3 2 1

*This volume is dedicated to
the memory of
Chih-Han Sah*

Contents

Preface

This volume continues the tradition of one of the most influential mathematical seminars of this century. As in previous books, this one "cuts across different subject areas" and reflects a variety of approaches to similar problems.

One of the main topics in the seminar for the past several years has been noncommutative algebra and geometry and its relations to modern physics. Different groups working in this area often use different methods. This book focuses on the program of one such group, led by M. Kontsevich and A. Rosenberg. Their approach to noncommutative smooth spaces, as reflected in their articles *Noncommutative Smooth Spaces* and *The Existence of Fiber Functors*, is based on the modern theory of categories.

The book is dedicated to the memory of a prominent colleague and close friend, Chih-Han Sah, and contains a review of of his work (joint with J. Dupont) entitled *Three Questions about Simplices in Spherical and Hyperbolic 3-Space*. A paper by T.V. Alekseyevskaya, A. Borovik, I.M. Gelfand, and N. White on *Matroid Homology* is dedicated to another type of geometry, namely, combinatorial geometry, while a paper by G. Cherlin, *Sporadic Homogeneous Structures* may be considered as a bridge between logic and geometry. Of course, there are always articles about Lie groups, algebras, and their representations. See the papers by V. Kac and A. Radul, *Poisson Structure for Restricted Lie Algebras* and A. Kazarnovski-Krol, *A Cycle for Integration Yielding the Zonal Spherical Function of Type A_n*.

I.M. Gelfand
V.S. Retakh
Rutgers University
October, 1999

The Gelfand Mathematical Seminars, 1996–1999

Matroid Homology

Tatiana V. Alekseyevskaya, Alexandre V. Borovik,[*]

I. M. Gelfand, and Neil White[†]

Abstract

We construct chain complexes of matroids and Lagrangian symplectic matroids which generalize Kontsevich's graph homology.

Introduction

The aim of the paper is to extend Kontsevich's definition of graph homology [K1, K2] to matroids and Lagrangian symplectic matroids. These concepts will be used in [BG] to develop a version of homology of fat graphs in terms of Lagrangian matroids. See [Mov] for another approach to homology of fat graphs.

1   Kontsevich graph homology

We work with finite non-oriented graphs, possibly with loops and multiple edges. An isomorphism between two graphs $\mathcal{G}$ and $\mathcal{G}'$ is a bijection $\alpha : E(\mathcal{G}) \longrightarrow E(\mathcal{G}')$ of their sets of edges which preserves the graph structure. A *loop* is an edge that connects a vertex with itself; an *isthmus* is an edge that does not belong to a cycle. Deletion of an isthmus from the graph increases the number of its connected components.

Consider the set of all triples $(\mathcal{G}, E, \leq)$ where $\mathcal{G}$ is a graph with a finite set of edges E, $n \geq 1$, and $\leq$ is a linear ordering of E. Abusing notation, we shall write $\mathcal{G}$ instead of $(\mathcal{G}, E, \leq)$. Consider the $\mathbb{Z}$-module $\widehat{\Gamma}_n$ freely

---

[*]The second author worked on this paper during his visits to the University of Florida, Gainesville, and Rutgers University.

[†]This work was supported in part by NSA grant MDA904-95-1-1056.

generated by all graphs $\mathcal{G}$ on ordered sets of n elements. Set $\widehat{\Gamma}_0 = 0$ and

$$\widehat{\Gamma} = \bigoplus_{n=0}^{\infty} \widehat{\Gamma}_n.$$

We define the differentials ∂_d (*deletion*) and ∂_c (*contraction*) by the following rules. If $\mathcal{G}$ is a graph on an ordered set $E = \{\, e_1 < \cdots < e_n \,\}$, then

$$\partial_c : \mathcal{G} \mapsto \sum (-1)^i \cdot \mathcal{G}/e_i$$

where the sum is taken over the set of all elements e_i of E which are not loops in $\mathcal{G}$, and

$$\partial_d : \mathcal{G} \mapsto \sum (-1)^i \cdot \mathcal{G} \setminus e_i$$

with the sum taken over the set of all elements e_i of E which are not isthmuses.

Lemma 1 *In this notation,*

$$\partial_c^2 = \partial_d^2 = 0$$

and

$$\partial_c \partial_d + \partial_d \partial_c = 0.$$

Therefore $(\widehat{\Gamma}, \partial_c)$ *and* $(\widehat{\Gamma}, \partial_d)$ *are chain complexes.*

Proof. The proof is the usual straightforward check, together with the observations that if e_i and e_j are neither ithmuses nor loops in $\mathcal{G}$, then e_i is a loop in $\mathcal{G}/e_j$ if and only if e_j is a loop in $\mathcal{G}/e_i$, e_i is an isthmus in $\mathcal{G} \setminus e_j$ if and only if e_j is an isthmus in $\mathcal{G} \setminus e_i$, and e_i cannot be a loop in $\mathcal{G} \setminus e_j$, nor an isthmus in $\mathcal{G}/e_j$. □

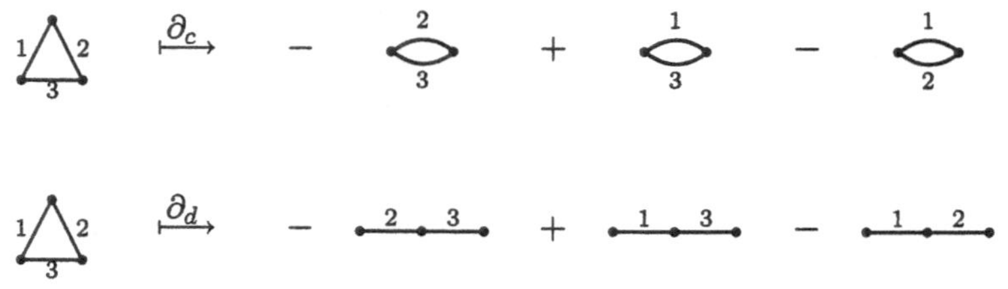

Figure 1: *Action of the differentials ∂_c and ∂_d on a graph.*

Kontsevich homology. If $(E, \leq)$ and $(F, \preceq)$ are two finite linearly ordered sets, the *relative sign* of a bijection $\alpha : E \longrightarrow F$ is defined in a natural way: we write E in the increasing order, $e_1 < \cdots < e_n$, and find the number p of inversions in the sequence

$$\alpha(e_1), \alpha(e_2), \ldots, \alpha(e_n);$$

then

$$\text{sign } \alpha = (-1)^p.$$

Let Γ_n be the factor of $\widehat{\Gamma}_n$ modulo the following relations: if $\mathcal{G}$ and $\mathcal{G}'$ are graphs isomorphic by virtue of a bijection of edges $\alpha : E \longrightarrow E'$, then

$$\mathcal{G}' = \text{sign } \alpha \cdot \mathcal{G}.$$

Thus

$$\Gamma_n = \bigoplus \mathbb{Z}\mathcal{G},$$

where the sum is taken over representatives of all isomorphisms classes of graphs with n edges. Note that if a graph $\mathcal{G}$ admits an automorphism that acts on its set of edges as an odd permutation, then $\mathcal{G} = -1 \cdot \mathcal{G}$ and $\mathbb{Z}\mathcal{G} \simeq \mathbb{Z}/2\mathbb{Z}$.

Set

$$\Gamma = \bigoplus_{i=0}^{\infty} \Gamma_n.$$

Denote by $\pi_n : \widehat{\Gamma}_n \longrightarrow \Gamma_n$ the canonical projection, so that $\pi = \bigoplus \pi_n$ is a homomorphism of graded modules $\pi : \widehat{\Gamma} \longrightarrow \Gamma$.

Theorem 2 *In the above notation, the differentials ∂_c and ∂_d map $\ker \pi_n$ into $\ker \pi_{n-1}$, $n = 1, 2, \ldots$. Hence they induce differentials denoted by the same symbols ∂_c and ∂_d) on the graded module Γ, turning it into algebraic chain complexes (Γ, ∂_c) and (Γ, ∂_d). Furthermore,*

$$\partial_c\partial_d + \partial_d\partial_c = 0.$$

Proof. We consider only the case of contraction; deletion can be treated in a similar way. It is enough to prove that if $\mathcal{G} = \mathcal{G}'$, then $\partial_c(\mathcal{G}) = \partial_c(\mathcal{G}')$.

Let $\mathcal{G}$ and $\mathcal{G}'$ be two graphs that are isomorphic by virtue of a bijection $\alpha : E \longrightarrow E'$ of their sets of edges. Let $e_1 < \cdots < e_n$ be the ordering of E. Denote by $o(i)$ the ordinal number of the element $\alpha(e_i)$ in the ordering $e'_1 \prec \cdots \prec e'_n$ of E'. Then

$$\partial_c(\mathcal{G}) = \sum(-1)^i \cdot \mathcal{G}/e_i \tag{1}$$

and

$$\partial_c(\mathcal{G}') = \sum (-1)^{o(i)} \cdot \mathcal{G}'/\alpha(e_i), \tag{2}$$

where in both cases the sum is taken over those indexes i for which the edges e_i (or, equivalently, $\alpha(e_i)$) are not loops. It is easy to check that

$$\operatorname{sign} \alpha|_{E \backslash \{e_i\}} = (-1)^{i+o(i)} \cdot \operatorname{sign} \alpha. \tag{3}$$

Therefore if the signs $(-1)^i$ and $(-1)^{o(i)}$ of the contractions $(-1)^i \cdot \mathcal{G}/e_i$ and $(-1)^{o(i)} \cdot \mathcal{G}'/\alpha(e_i)$ in Equations 1 and 2 are different, then the numbers i and $o(i)$ have different parity, and by Equation 3,

$$\mathcal{G}/e_i = -1 \cdot \mathcal{G}'/\alpha(e_i)$$

and eventually

$$(-1)^i \cdot \mathcal{G}/e_i = (-1)^{o(i)} \cdot \mathcal{G}'/\alpha(e_i).$$

The case where i and $o(i)$ have the same parity similarly works, and this proves $\partial_c(\mathcal{G}) = \partial_c(\mathcal{G}')$. $\square$

The homology of the algebraic chain complex (Γ, ∂_c) is called the *Kontsevich homology*.

2   Matroid homology

Matroids. Now we wish to transfer the definition of Kontsevich homology to the case of matroids. Recall that a matroid on a finite set E is a set $\mathcal{M}$ of subsets (called *bases*) of E that satisfies the following Exchange Axiom:

for any $A, B \in \mathcal{M}$ and $a \in A \backslash B$ there is $b \in B \backslash A$ such that $A \cup \{b\} \backslash \{a\}$ belongs to $\mathcal{M}$.

It can be shown that all bases of a matroid contain the same number of elements; this number is called the *rank* of $\mathcal{M}$. The *dual matroid* $\mathcal{M}^*$ is defined as the set

$$\mathcal{M}^* = \{ E \backslash A \mid A \in \mathcal{M} \};$$

it can be checked that $\mathcal{M}^*$ is a matroid. Obviously, $(\mathcal{M}^*)^* = \mathcal{M}$. It is well-known that if E is the set of edges of a graph $\mathcal{G}$, then the set $\mathcal{M}$ of all maximal forests in $\mathcal{G}$ is a matroid on E.

Deletion and contraction. The operations of contraction of an edge and deletion of an edge in a graph have natural matroid analogues.

An element $e \in E$ is called a *loop* in a matroid $\mathcal{M}$, if e does not belong to any basis of $\mathcal{M}$; e is an *isthmus* if e is contained in every basis of $\mathcal{M}$.

If $\mathcal{M}$ is a matroid of rank k on a set E and $e \in E$ is not an isthmus, then the *deletion* $\mathcal{M} \setminus e$ is the matroid on $E \setminus \{e\}$ whose bases are all bases of $\mathcal{M}$ that do not contain e. If e is not a loop, the *contraction* $\mathcal{M}/e$ is the matroid on $E \setminus \{e\}$ defined as

$$\mathcal{M}/e = \{ A \setminus \{e\} \mid A \in \mathcal{M} \text{ and } e \in A \}.$$

Notice that deletion of an element e from a matroid is equivalent to contraction of e in the dual matroid, and vice versa:

$$\mathcal{M}^* / e = (\mathcal{M} \setminus e)^*, \quad \mathcal{M}^* \setminus e = (\mathcal{M}/e)^*.$$

Furthermore, if $\mathcal{M}$ is the matroid of maximal forests of the graph $\mathcal{G}$, the deletion $\mathcal{M} \setminus e$ and contraction $\mathcal{M}/e$ of $\mathcal{M}$ are exactly the matroids of maximal forests of the graphs $\mathcal{G} \setminus e$ and $\mathcal{G}/e$.

The chain complex. Consider the set of all triples $(\mathcal{M}, E, \leq)$, where $\mathcal{M}$ is a matroid on a finite set E of cardinality n, $n \geq 1$, and $\leq$ is a linear ordering of E. Abusing notation, we shall write $\mathcal{M}$ instead of $(\mathcal{M}, E, \leq)$. Consider the $\mathbb{Z}$-module $\widehat{M}_n$, generated by all matroids on ordered sets of n elements. Set $\widehat{M}_0 = 0$ and

$$\widehat{M} = \bigoplus_{n=0}^{\infty} \widehat{M}_n.$$

Notice that the map $* : \mathcal{M} \mapsto \mathcal{M}^*$, which sends every matroid to its dual, extends to isomorphism $* : \widehat{M} \longrightarrow \widehat{M}$.

We define the differentials ∂_d (*deletion*) and ∂_c (*contraction*) by analogy with graphs: if $\mathcal{M}$ is a matroid on a set E ordered $e_1 < \cdots < e_n$, then

$$\partial_c : \mathcal{M} \mapsto \sum (-1)^i \cdot \mathcal{M}/e_i$$

where the sum is taken over the set of all elements e_i of E that are not loops in $\mathcal{M}$, and

$$\partial_d : \mathcal{M} \mapsto \sum (-1)^i \cdot \mathcal{M} \setminus e_i,$$

with the sum taken over the set of all elements e_i of E that are not isthmuses.

Lemma 3 *In the notation above,*

$$\partial_c^2 = \partial_d^2 = \partial_c\partial_d + \partial_d\partial_c = 0.$$

The proof is entirely analogous to that for graphs.

Let now M_n be the factor of $\widehat{M}_n$, modulo the relations defined by the following rule: if $\mathcal{M}$ and $\mathcal{M}'$ are matroids isomorphic by virtue of a bijection
$\alpha : E \longrightarrow E'$, then $\mathcal{M}' = \operatorname{sign}\alpha \cdot \mathcal{M}$. Thus

$$M_n = \bigoplus \mathbb{Z}\mathcal{M}$$

where the sum is taken over all representatives of isomorphisms classes of matroids on n elements. Note that $\mathbb{Z}\mathcal{M} \simeq \mathbb{Z}/2\mathbb{Z}$ or $\mathbb{Z}$ depending on whether or not the matroid $\mathcal{M}$ admits an automorhism which acts on its set E of elements as an odd permutation. Set

$$M = \bigoplus_{n=o}^{\infty} M_n,$$

and denote by π the canonical homomorphism of graded modules $\pi :$ $\widehat{M} \longrightarrow M$, with components

$$\pi_n : \widehat{M}_n \longrightarrow M_n.$$

Theorem 4 *The differentials ∂_c and ∂_d map $\ker\pi_n$ into $\ker\pi_{n-1}$, $n = 1, 2, \ldots$. In particular, ∂_c and ∂_d induce differentials (which we denote by the same symbols) on the graded module M, turning it into algebraic chain complexes (M, ∂_c) and (M, ∂_c). Furthermore,*

$$\partial_c\partial_d + \partial_d\partial_c = 0.$$

The proof is entirely analogous to that of graphs.

We shall call the homology of the chain complex (M, ∂_c) the *matroid homology*.

3 Lagrangian matroid homology

Lagrangian matroids. Following [BGW2], consider the set

$$[n] \sqcup [n]^* = \{\,1, \ldots, n\,\} \sqcup \{\,1^*, \ldots, n^*\,\}$$

with the involutive map

$$* : i \mapsto i^*, \ i^* \mapsto i.$$

A subset $K \subseteq [n] \sqcup [n]^*$ is called *admissible* if $K \cap K^* = \emptyset$. A linear ordering $\prec$ on $[n] \sqcup [n]^*$ is *admissible* if $i \prec j$ implies $j^* \prec i^*$. The *standard* admissible ordering on $[n] \sqcup [n]^*$ is

$$n^* < \cdots < 1^* < 1 < \cdots < n.$$

A permutation w of the set $[n] \sqcup [n]^*$ is *admissible* if it commutes with $*$, that is, $w(i^*) = (w(i))^*$. The group of all admissible permutations of $[n] \sqcup [n]^*$ is the Coxeter group BC_n, also known as the *hyperoctahedral group*.

For every admissible ordering $\prec$ of $[n] \sqcup [n]^*$, we define the ordering on the set $\mathcal{A}_k$ of admissible k-element subsets as follows: if

$$A = \{ i_1 \prec \cdots \prec i_k \} \quad \text{and} \quad B = \{ j_1 \prec \cdots \prec j_k \}$$

then

$$A \prec B \quad \text{iff} \quad i_1 \prec j_1, \ldots, i_k \prec j_k.$$

We say that a subset $\mathcal{M} \subseteq \mathcal{A}_k$ is a *symplectic matroid* if it satisfies the *Maximality Property*: for every admissible ordering $\prec$, the set $\mathcal{M}$ contains a unique maximal element A: $B \prec A$ for all $B \in \mathcal{M}$. Elements of $\mathcal{M}$ are called *bases*, k is the *rank* of $\mathcal{M}$. A symplectic matroid on $[n] \sqcup [n]^*$ is called *Lagrangian* if it has rank n. Lagrangian matroids have been also introduced by Bouchet [Bou], under the name of *symmetric matroids*; this concept is equivalent to that of a Δ-*matroid* and is almost the same as *metroid* of Dress and Havel [BDH, DH]. The equivalence of the definitions of Δ-matroids and symplectic Lagrangian matroids is shown by Wenzel [W].

Let $\mathcal{L}$ be a Lagrangian matroid. Consider the Euclidean real vector space $\mathbb{R}^n$ with the canonical orthonormal basis $\epsilon_1, \ldots, \epsilon_n$ and the coordinates $x_1, \ldots, x_n$. It will be convenient to set $\epsilon_{i^*} = -\epsilon_i$ for $i \in [n]$. Then the hyperoctahedral group $W = BC_n$ acts on $\mathbb{R}^n$ by orthogonal transformations $w \cdot \epsilon_i = \epsilon_{w \cdot i}$.

Take the convex hull Δ of the points δ_B, defined for each basis $B \in \mathcal{L}$ following the rule that

$$\delta_B = \sum_{i \in B} \epsilon_i.$$

Then δ_B are vertices of Δ and also vertices of the n-cube $[1, -1]^n$ in $\mathbb{R}^n$. Δ is a Coxeter matroid polytope in the sense of [BGW1, GS]: its edges are parallel to roots in the root system

$$C_n = \{ \pm\epsilon_i \pm \epsilon_j \neq 0 \mid i, j \in [n] \}.$$

Interpretation of deletion and contraction in terms of Lagrangian matroids. Now if $\mathcal{M}$ is an (ordinary) matroid on the set $[n]$, take, for every basis $B \in \mathcal{M}$, the set $\overline{B} = B \cup ([n] \setminus B)^*$; all these sets form a Lagrangian matroid $\overline{\mathcal{M}}$ on $[n] \sqcup [n]^*$ [BGW2].

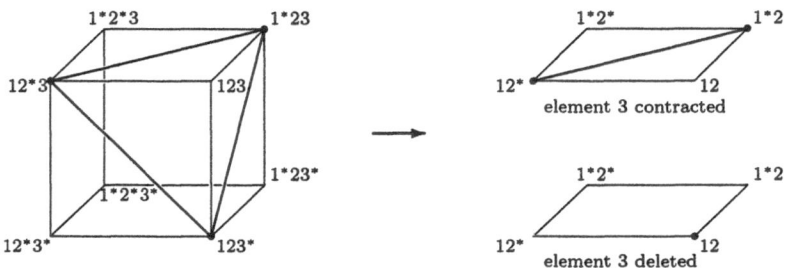

Figure 2: *Effects of contraction and deletion in a matroid on its matroid polytope.*

Let $\Delta = \Delta(\mathcal{M})$ be the matroid polytope for $\overline{\mathcal{M}}$; it retains evidence of its ordinary matroid origin: the edges of Δ are parallel to the roots of the root system

$$A_{n-1} = \{\, \epsilon_i - \epsilon_j \mid i,j \in [n], i \neq j \,\}$$

canonically embedded into the root system system C_n [BGW2].

Mark the facet $x_i = 1$ of the cube by i, and the facet $x_i = -1$ by i^*. Then deletion of the element i from the matroid $\mathcal{M}$ means taking the intersection of Δ with the facet i^* of the n-cube $[1, -1]^n$; contraction of i means intersecting Δ with the facet i of the cube. Figure 2 illustrates these operations for the matroid with the bases 12, 23, 13 on [3], which is exactly the matroid of maximal forests in the triangular graph of Figure 1. Notice that if i is a loop (resp. isthmus) in $\mathcal{M}$, then the intersection of Δ with the facet i (resp. i^*) is empty.

This suggests the definition of contraction for Lagrangian matroids. If an element $i \in [n] \sqcup [n]^*$ is not a *loop* in a Lagrangian matroid $\mathcal{L}$, i.e, if i lies in at least one basis of $\mathcal{L}$, then the set

$$\mathcal{L}' = \{\, B \setminus \{i\} \mid B \in \mathcal{L} \text{ and } i \in B \,\}$$

is a Lagrangian matroid on $([n] \sqcup [n]^*) \setminus \{i, i^*\}$. This can be seen from the observation that its matroid polytope Δ' is the intersection of the polytope Δ for $\mathcal{L}$ with the facet i of the cube $[1, -1]^n$; hence Δ' is a face of Δ and the edges of Δ' are parallel to the roots in the root system C_n. After that [BGW2] asserts that $\mathcal{L}$ is a Lagrangian symplectic matroid.

We shall call $\mathcal{L}'$ the *contraction* of element i in $\mathcal{L}$ and write $\mathcal{L}' = \mathcal{L} / i$. If i is a loop on $\mathcal{L}$, then $\mathcal{L} / i$ is the empty set.

Double sets.   A *double set* is a pair $(E, *)$ where E is a set of $2n$ elements and $*$ an involutive permutation of E without fixed points. A morphism

$$(E, *) \longrightarrow (E', *)$$

of double sets is a injection of E onto E', which preserves the action of involution $*$. An *admissible ordering* of a double set $(E, *)$ is a bijective morphism

$$\alpha : [n] \sqcup [n]^* \longrightarrow (E, *);$$

if we denote $e_i = \alpha(i)$ for $i \in [n] \sqcup [n]^*$, then E can be written as an ordered set

$$E = \{ e_{n^*} \prec \cdots \prec e_{1^*} \prec e_1 \prec \cdots \prec e_n \},$$

so that f transfers to E the standard ordering from $[n] \sqcup [n]^*$. Definitions of admissible sets, symplectic and Lagrangian matroids can be transferred, in a very obvious way, from $[n] \sqcup [n]^*$ to any double set E: a set $\mathcal{L}$ of admissible k-sets of a double set E is a symplectic matroid if, for any admissible ordering of E, $\mathcal{L}$ contains a unique maximal element. A bijective morphism $\alpha : E \longrightarrow E'$ of a double set is an *isomorphism* of symplectic matroids $\mathcal{L}$ and $\mathcal{L}'$ on E and E', correspondingly, if α induces a bijective map of $\mathcal{L}$ onto $\mathcal{L}'$.

If

$$\alpha : [n] \sqcup [n]^* \longrightarrow (E, *) \text{ and } \beta : [n] \sqcup [n]^* \longrightarrow (E', *)$$

are two ordered double sets and $\gamma : (E, *) \longrightarrow (E', *)$ a morphism, the *relative sign* is defined as

$$\text{sign}\, \gamma = \det(\beta^{-1} \circ \gamma \circ \alpha),$$

where the admissible permutation $\beta^{-1} \circ \gamma \circ \alpha$ of $[n] \sqcup [n]^*$ is treated as an orthogonal transformation of $\mathbb{R}^n$.

The graded $\mathbb{Z}$-module of Lagrangian matroids.   We simply repeat, with obvious modification, the definition of the chain complex of matroids given in the previous section.

Consider the $\mathbb{Z}$-module $\hat{\Lambda}_n$ generated by all Lagrangian matroids $\mathcal{L}$ on ordered double sets of cardinality $2n$. Set $\hat{\Lambda}_0 = 0$ and

$$\hat{\Lambda} = \bigoplus_{n=0}^{\infty} \hat{\Lambda}_n.$$

It would be convenient to set $\mathcal{L} = 0$ for the empty Lagrangian matroid.

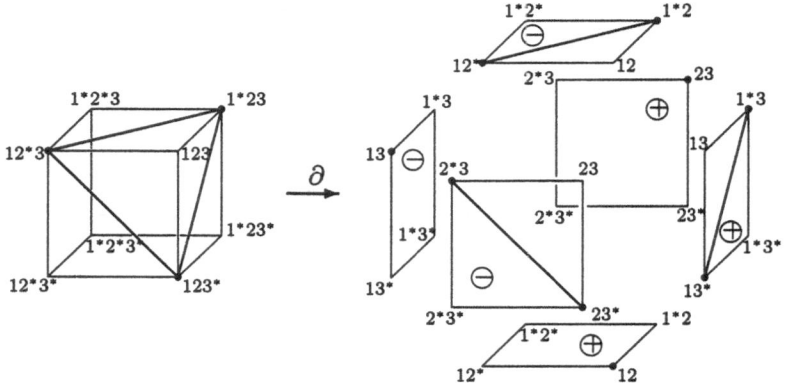

Figure 3: Action of the differential ∂ on the Lagrangian matroid { 123*, 12*3, 1*23 } which corresponds to the graph and the matroid of the previous examples, Figures 1 and 2. The signs on the facets are the signs in Equation 4.

Differential on $\hat{\Lambda}$. For $i \in [n]$, define $(-1)^{i^*} = -(-1)^i$. We define the differential ∂ on $\hat{\Lambda}$ by

$$\partial : \mathcal{L} \mapsto \sum_{i \in [n] \sqcup [n]^*} (-1)^i \mathcal{L} / e_i \qquad (4)$$

Lemma 5 *In the notation above,*

$$\partial^2 = 0$$

and $(\hat{\Lambda}, \partial)$ is an algebraic chain complex.

Again, the proof is a straightforward check, noticing that e_i is a loop of $\mathcal{L}/e_j$ if and only if e_j is a loop of $\mathcal{L}/e_i$.

Now let Λ_n be the factor of $\hat{\Lambda}_n$, modulo the relation which is defined exactly as in the case of graphs or matroids: if $\mathcal{L}$ and $\mathcal{L}'$ are Lagrangian matroids isomorphic by virtue of a bijection $\alpha : E \longrightarrow E'$ of their underlying double sets, then $\mathcal{L}' = \operatorname{sign} \alpha \cdot \mathcal{L}$.

Thus

$$\Lambda_n = \bigoplus \mathbb{Z}\mathcal{L}$$

where the sum is taken over all representatives of isomorphisms classes of Lagrangian matroids on $2n$ elements. Note that $\mathbb{Z}\mathcal{L} \simeq \mathbb{Z}/2\mathbb{Z}$ or $\mathbb{Z}$

depending on whether the Lagrangian matroid $\mathcal{L}$ admits an automorhism of sign -1, or not.

Finally, set $\Lambda_0 = 0$ and

$$\Lambda = \bigoplus_{n=0}^{\infty} \Lambda_n.$$

Denote by π_n the natural projection $\pi_n : \hat{\Lambda}_n \longrightarrow \Lambda_n$ and form $\pi = \bigoplus \pi_n$.

Theorem 6 *The differential ∂ maps $\ker \pi_n$ into $\ker \pi_{n-1}$, $n = 1, 2, \ldots$. In particular, ∂ induces a differential (which we denote by the same symbol ∂) on the graded module Λ, turning it into an algebraic chain complex (Λ, ∂).*

Proof. The proof is analogous to that for graphs or matroids, with the only exception being that instead of Equation 3, one has to use the following observation.

Lemma 7 *Let $\alpha : E \longrightarrow E'$ be a bijective morphism of ordered double sets*

$$E = \{ e_{n^{\bullet}} < \cdots < e_{1^{\bullet}} < e_1 < \cdots < e_n \}$$

and

$$E' = \{ e'_{n^{\bullet}} \prec \cdots \prec e'_{1^{\bullet}} \prec e'_1 \prec \cdots \prec e'_n \}.$$

If $i \in [n] \sqcup [n]^*$ and $\alpha(e_i) = e'_j$, then

$$\operatorname{sign} \alpha|_{E \setminus \{ i, i^* \}} = (-1)^i \cdot (-1)^j \cdot \operatorname{sign} \alpha. \tag{5}$$

$\square$

4   Some chain maps

If $\mathcal{G}$ is a graph with the set of edges $E = E(\mathcal{G})$, the collection $\mathcal{T}$ of maximal forests in $\mathcal{G}$ is a (collection of bases of a) matroid on the set E. This extends to the homomorphism of graded $\mathbb{Z}$-modules

$$\tau : \Gamma \longrightarrow M.$$

Theorem 8 *The homomorphism τ is a chain map of the chain complexes*

$$\tau : (\Gamma, \partial_c) \longrightarrow (M, \partial_c)$$

and

$$\tau : (\Gamma, \partial_d) \longrightarrow (M, \partial_d),$$

that is,

$$\tau\partial_c = \partial_c\tau \quad and \quad \tau\partial_d = \partial_d\tau.$$

Analogously, the map λ that assigns to an ordinary matroid $\mathcal{M}$ on $[n]$ the homogeneous Lagrangian matroid $\overline{\mathcal{M}}$ extends to a homomorphism of graded $\mathbb{Z}$-modules

$$\lambda : M \longrightarrow \Lambda.$$

Theorem 9 *The map λ satisfies the identity*

$$\partial\lambda = \lambda(\partial_c - \partial_d).$$

Bicomplex of matroids. Notice that contraction of an element in a matroid $\mathcal{M}$ decreases its rank by 1 and deletion of an element decreases by 1 the rank of the dual matroid $\mathcal{M}^*$. Consider the decomposition

$$M_n = \bigoplus_{k=0}^{n} M_{k,n-k}$$

of M_n into the direct sum of modules $M_{k,n-k}$ generated by matroids $\mathcal{M}$ of rank k on n elements, and set $M_{k,n-k} = 0$ if $k < 0$ or $n - k < 0$. Then the differentials ∂_c and ∂_d act

$$\partial_c : M_{k,n-k} \longrightarrow M_{k-1,n-k}, \qquad \partial_d : M_{k,n-k} \longrightarrow M_{k,n-k-1}$$

and the family $\{ M_{k,n-k} \}$ has a natural structure of a bicomplex with the differentials ∂_c and $-\partial_d$ [M, Ch. XI, §6]. In this interpretation the map $\lambda : M \longrightarrow \Lambda$ from the chain complex of ordinary matroids to the chain complex of symplectic Lagrangian matroids can be viewed as the *totalisation* of the bicomplex $(\{ M_{k,n-k} \}, \partial_c, -\partial_d)$.

References

[BG]   A. V. Borovik, I. M. Gelfand, Homology of fat graphs, in preparation.

[BGW1] A. V. Borovik, I. Gelfand and N. White, Coxeter matroid polytopes, to appear, *J. Alg. Comb.*

[BGW2] A. V. Borovik, I. Gelfand and N. White, Symplectic matroids, to appear, *J. Alg. Comb.*

[Bou] A. Bouchet, Greedy algorithm and symmetric matroids, *Mathematical Programming* **38** (1987), 147–159.

[BDH] A. Bouchet, A. Dress and T. Havel, Δ-matroids and metroids, *Adv. Math.* **91** (1992), 136–142.

[DH] A. Dress and T. Havel, Some combinatorial properties of discriminants in metric vector spaces, *Adv. Math.* **62** (1986), 285–312.

[GS] I. Gelfand and V. V. Serganova, Combinatorial geometries and torus strata on homogeneous compact manifolds, *Russian Math. Surveys* **42** (1987), 133–168; see also I. M. Gelfand, *Collected Papers*, vol. III, Springer-Verlag, New York a.o., 1989, pp. 926–958.

[K1] M. Kontsevich, Formal (non)-commutative symplectic geometry, in *The Gelfand Mathematical Seminars, 1990–1992*, Birkhäuser, 1993, pp. 173–187.

[K2] M. Kontsevich, Feynman diagrams and low-dimensional topology, in *First European Congress of Mathematics, Paris, July 6–10, 1992*, Birkhäuser, 1994, vol. 2, pp. 97–122.

[M] S. Maclane, *Homology*, Springer-Verlag, 1963.

[Mov] M. Movshev, The definition of graph homology of algebras, preprint.

[W] W. Wenzel, *Geometric Algebra of Δ-matroids and Related Combinatorial Geometries*, Habilitationschrift, Bielefeld, 1991.

T. V. Alekseyevskaya
Center for Mathematics, Science
and Computer Education
Rutgers University
New Brunswick, NJ 08903

A. V. Borovik
Department of Mathematics
UMIST
PO Box 88
Manchester M60 1QD, UK

I. M. Gelfand
Department of Mathematics
Rutgers University
New Brunswick, NJ 08903

N. White
Department of Mathematics
University of Florida
Gainesville, FL 32605

AMS Subject Classification: 05B35

Sporadic Homogeneous Structures

*Gregory Cherlin**

Abstract

We discuss Lachlan's classification theory for *finite homogeneous structures* and related problems on finite permutation groups. Lachlan's theory provides a hierarchy of classifications in which structures which are "sporadic" in one context reappear as members of infinite families at later stages. Every finite structure is accounted for at some level in this hierarchy, but for structures associated with familiar primitive permutation groups, the combinatorial problem of locating that level precisely can be quite challenging.

Key words: classification, sporadic, homogeneity, permutation group, stability

> . . .   *in most categories few objects have the Witt property; those that do are very well behaved indeed.*
> [As, p. 82]

Introduction

When classification results are enlivened by the appearance of uninvited guests in the form of "sporadic" objects, those who take an interest in these interlopers may be tempted to account for them in various ways, possibly by viewing them as coming from infinite (perhaps even continuous) families of more general objects which may be natural from some broader point of view. In pure model theory, Lachlan's classification theory for finite homogeneous relational structures provides a relatively well understood illustration (or "toy model") of this sort of thing. This theory, which will be reviewed below, provides an infinite number of classification

*Supported in part by the Binational Science Foundation BSF 0377215.

theorems of a general character for combinatorial structures with rich automorphism groups, parametrized by certain bounds on the complexity of the structures. Any finite structure will actually appear at some stage in one of these classifications, and may well occur as a sporadic structure initially; in the long run, every sporadic structure winds up belonging to a family parametrized by numerical invariants; at any given stage, only finitely many structures occur as sporadics; finally, one will never "move beyond" the sporadics: we will always encounter new structures making their appearance as (temporarily) sporadic structures.

That this sort of thing would occur on a regular basis is only natural, although it may be surprising that this state of affairs would be described by a theorem in a natural and reasonably general setting. The general theorem is actually a consequence of one single finiteness result – the finiteness of the set of sporadic simple groups – and some permutation group theory. (See §2 for an explanation of how we can view structures as a special case of groups, rather than the other way around.) It should be said that whatever the motivation, in working out this theory one does not need to think about the phenomenon of sporadicity as such, and what is really involved is the finiteness theorem given below as a Coordinatization Theorem (§5), which is equivalent to the Bounded-Rank Theorem mentioned briefly in §7.

The notion of homogeneity also leads one to associate a natural measure of complexity $\kappa(G, X)$ with any finite permutation group (G, X), which may be of interest in its own right; this can be studied from a combinatorial point of view without reference to any theoretical background in model theory. This invariant behaves somewhat like a dimension; for example, for a vector space of dimension n (that is, for the group $(\mathrm{GL}(V), V)$) κ will be $n + 1$. Determining the complexity of specific permutation groups with precision can be quite challenging, and we have included a selection of open problems in the final section, which can be read independently of the presentation of the general theory, although it will no doubt be helpful to look over the background material in the first three sections, which contain a number of specific examples.

In the last example of §3 we will see how one fairly rich family of examples divides into families and sporadics at each level of analysis, with the number of families and sporadics finite at each stage; but in this case at least, the number of structures counted as sporadic is exponentially large compared to the number of parametrized families encountered.

Pure model theory at the present time consists largely of ideas connected with Shelah's "classification theory," which attempts to provide a very general theory of classification of structures. This comes in a number of variants, most of which emphasize the classification of infinite structures. For example, when these ideas are specialized to the case of modules over a ring, they involve the classification of pure-injective indecomposable modules, with ideas very closely related to representation theory. It was seen by Lachlan that the model theoretic approach also makes sense for certain broad classes of finite structures: homogeneous structures (for a finite relational language). These results have since been generalized (notably by Hrushovski) to cover reasonably broad classes of finite permutation groups. We will say a little about this as well, mainly in §8.

The point of Lachlan's theory is that it involves an infinite number of related classification problems; for each type of structure, the homogeneous structures of that type can be rather thoroughly classified. In each instance, the structures involved fall naturally into a finite number of families, and within each family the individual objects are parametrized by a finite number of numerical invariants which may be varied independently. It is possible for the number of invariants needed to describe an object to be *zero*, in which case the family degenerates to a single structure, and these are the structures which may be considered *sporadic*. We have said that *any* finite object will eventually be covered by one of these classification schemes; in other words, any structure can be viewed as homogeneous of some type (see the example in §1 and the general considerations of §2). In passing from one classification problem to the next, what generally occurs is that (1) new families arise; (2) in the old families, additional numerical invariants are acquired which may be varied independently. In particular, the sporadic objects from one classification scheme are eventually absorbed into parametrized families. So in this model, sporadic objects can always be understood as part of some larger classification scheme, but sporadicity itself is not escaped.

In the next section we will give a concrete example of all of this: a 1-parameter family of graphs which contains three homogeneous graphs but which enters as a family of homogeneous structures of a slightly more complex type. We will see later that a slight generalization of this provides simple and very uniform families of graphs, some of which will be considered sporadic at more or less every level of Lachlan's hierarchy. The analysis of these examples has been carried out with extraordinary precision by Saracino, building on the rough analysis of [CMS]. The bulk of his results are summarized at the end of §3.

Although Lachlan's theory provides a classification of the homogeneous finite relational structures of any specified type (for definitions, see §2), the literature generally refers to the classification of homogeneous *stable* relational structures, a broader class with a comparatively technical definition, which turns out to consist of the finite ones and their infinite limits. There are good reasons for this generalization: to prove the theorems, it is very convenient to move back and forth between the large finite structures and their infinite limits. Shelah's notion of stability is one of the fundamental concepts of pure model theory; Lachlan realized that in the context of homogeneous relational systems, it is equivalent to "smooth limit of finite" (cf. §4). This has led to a fruitful interaction of model theory and the theory of permutation groups, which involves an interplay between the group theoretic analysis of large finite structures, and the combinatorial analysis of their infinite limits. All of this depends ultimately on the classification of the finite simple groups.

There are a number of expositions of this theory. The theory of (and some major open problems concerning) homogeneous structures in general and finite or stable ones in particular is discussed in Lachlan's ICM lecture [La2]. A detailed account of the classification theory for finite homogeneous structures is given in [KL]; this combines a detailed exposition with some major expositions and an important advance (the form of the Stretching Theorem given in §6 below). It does assume familiarity with the language and point of view of model theory, which to some extent the present article is intended to provide (via examples, the general discussion in §2, and a handful of technical definitions). The subject of homogeneity is also the final topic taken up in the text [DM], which also discusses a number of other topics in permutation group theory which have been important to its users in model theory, such as the O'Nan-Scott Theorem.

In the long run, the most general form of this theory would be a structural analysis of large k-closed permutation groups of bounded rank (with both k and the bound on rank taken as fixed). This has not been carried out, but an intermediate stage, in which the bound on rank is sharpened to a bound on the number of orbits on 4-tuples, is discussed in [Hr]. This depends on a thorough analysis of the primitive case by group theoretic methods [KLM]; but the analysis of the general case requires a very heavy dose of model theory, and indeed traditional permutation group theory has little to say about the imprimitive case. Technicalities aside, the final form of the theory is quite similar to the theory we will encounter in Lachlan's original case, with the main difference being the appearance in [KLM] of structures conspicuous by their absence in Lachlan's theory: vector spaces with their classical adornments (inner products and quadratic forms), and – in characteristic 2 – also some less classical adornments (see §8).

I thank Saracino for keeping me apprised of his very delicate work, which contains the most subtle analysis of the behavior of the parameter κ (§2) for any specific family of primitive permutation groups.

1 The graphs n^2

In the next section we will present the general correspondence between finite permutation groups and finite structures, which are essentially the same thing. Here we look at one example of a family of structures or permutation groups whose properties illustrate some of the issues that turn out to be central in the general theory. This example can be analyzed easily in complete detail, but its natural generalization is still not thoroughly understood (§9, Problem 3).

Definition The graph n^d has as its vertex set the set of all d-sets taken from an n-element set, with edges between d-tuples u and v if they differ at exactly one vertex.

Note that the automorphism group of n^d is the wreath product $\text{Sym}(n) \wr \text{Sym}(d)$; for this reason this graph may be referred to variously as a "power" or a "wreath product action", according as one pays more attention to the graph or permutation group; these two points of view are equivalent for our purposes.

In the present section we consider only the graphs n^2, which are also referred to as the *line graphs* $L(K_{n,n})$, as they can be viewed as graphs derived from the complete bipartite graphs $K_{n,n}$ by taking the edges of $K_{n,n}$ as vertices, with two edges adjacent if they have a vertex in common in $K_{n,n}$. For $n \leq 3$ these graphs are *homogeneous* in the following sense: any isomorphism between two induced subgraphs is the restriction of some automorphism. For $n > 3$ they are not homogeneous as graphs, because they then contain two classes of graphs of the form $2 \cdot K_2$ (meaning, two disjoint edges): "parallel" edges and "orthogonal" edges.

For $n = 3$ the graph obtained is considered sporadic (either informally, or as an instance of Lachlan's theory). For $n = 1, 2$ these graphs are not sporadic. For $n = 2$ the graph is a complete bipartite graph and belongs to a family of homogeneous complete multipartite graphs parametrized by two numerical parameters, while for $n = 1$ the graph belongs to the complementary family: the complement of a complete multipartite graph is a disjoint union of complete graphs.

What matters here, though, is the fact that all the graphs n^2 do in fact occur as one infinite family of homogeneous structures – but not as *graphs*. We may instead consider these graphs as coming equipped with a 4-place relation of parallelism, $P(v_1, v_2, v_3, v_4)$. This may be defined from the edge relation as follows: if two disjoint edges have the property that the four vertices involved have no further edges between them, the edges are *parallel* if there is no vertex adjacent to all four of the given vertices.

In particular the automorphism group of the original graph is also the automorphism group of the enriched structure. However, in the category of graphs-with-parallelism, all of these structures are homogeneous, in the sense that parallelism-preserving isomorphisms between subgraphs are induced by automorphisms.

An important point here is that we consider the two structures on n^2 – the graph structure, or the structure of a graph with parallelism – as identical structures, even though they are of different types. We justified this by looking at them as permutation groups; logicians would have expressed this directly, in terms of the structures, by stating that the relations in each structure are definable from the relations in the other structure (definable over $\emptyset$, specifically). Most of the notions coming from model theory can be translated into the language of permutation groups, and it can be technically advantageous to do so on occasion (just as it can be equally advantageous, on other occasions, to make the translation in the other direction). In particular the very rich information contained in the classification of the finite simple groups is largely inaccessible from the structural point of view, though it looms very large indeed in the analysis of the associated permutation groups.

Thus if one traces through Lachlan's classification theory for the case of graphs, the graph 3^2 will necessarily occur sporadically (while 1^2 and 2^2 can easily be absorbed into other infinite families); once the classification is extended to a sufficiently rich class of 4-hypergraphs, the infinite family n^2 will occur as an infinite family indexed by the parameter n (this happens, so to speak, automatically, on the basis of general principles).

For the graphs n^d the situation is more complicated and as d increases the complexity of the additional relations which must be considered goes to infinity, as does the number of exceptional cases (with d large relative to n) which occur prematurely as sporadic examples. The place of these structures in the hierarchy of homogeneous structures was determined with extraordinary precision by Saracino. We will present this in §3 after setting up our point of view in general in the next section.

If one replaces the natural representation of $\mathrm{Sym}(n)$ on n elements by its representation on k-sets (so the original representation corresponds to $k = 1$) then the analysis remains very incomplete, though it is possible that Saracino's analysis extends in some reasonable way to $k > 1$.

2   Structures and permutation groups

The structures considered in Lachlan's theory will be *relational structures* $\mathcal{X} = (X; \mathbf{R})$ with $\mathbf{R} = (R_1, \ldots, R_r)$ a finite sequence of relations. If the relation R_i is an n_i-ary relation, the *signature* of the relational structure $\mathcal{X}$ is the sequence $\tau = (n_1, \ldots, n_r)$. Thus a signature is a finite sequence of natural numbers, and defines a certain type (or category) of relational structure. There are good reasons to add some more data to the signature (for example, one could impose generalized symmetry and irreflexivity properties on each relation) but laying this out in detail would be both tedious and irrelevant to our present purpose.

There is a Galois connection between finite structures and finite permutation groups. We assign to the structure $(\mathcal{X}, \mathbf{R})$ the permutation group (Aut $\mathcal{X}, X$). In the other direction, given a permutation group (G, X), a relation on X is called *invariant* (specifically, G-invariant, if the context is not otherwise clear) if it is invariant under the action of G. We may associate to a finite permutation group (G, X) the relational structure $\mathcal{X}$ whose relations are all the k-ary invariant relations for $k \leq |X|$.

This establishes a 1–1 correspondence between the image of the connection on both sides: in other words, between the faithful finite permutation groups and some structures, called *canonical structures*, not much encountered in nature, but much encountered in model theory: they are essentially the structures in which every definable relation is given explicitly as part of the structure (apart from the generous cut-off at complexity k; there is not, in any case, much use for $(k+1)$-ary relations on a set of size k). For many purposes finite structures that correspond to the same permutation group should be identifed; this was illustrated in the previous example by our observation that the parallelism relation is implicit in the graph structure on n^2, and hence might as well be added as an ingredient of the structure.

At any given moment one is likely to be working on one side or the other of this Galois connection, but using notions that originate on both sides; so it is useful to build up a glossary giving the meaning of concepts originating on one side in the language of the other. We will give some examples.

As we have said, a relation R on the set X is said to be *invariant* – or alternatively *definable* – in the structure $\mathcal{X} = (X, \mathbf{R})$ if it is (Aut $\mathcal{X}$)-invariant. This coincides with *first order definability* without parameters when X is finite – hence the alternative terminology.

A structure is *primitive* if it has no nontrivial invariant equivalence relation – this notion has always been emphasized in permutation group theory, and the same terminology has been adopted by model theorists.

A permutation group is k-*closed* if it is the image in the Galois connection of some structure all of whose relations have at most k places, and the k-*closure* of a permutation group is the smallest k-closed group containing it. The intrinsic definition runs as follows: the k-closure of (G, X) is the set of permutations σ with the property that for any k-tuple $\mathbf{x}$ in X, the image of $\mathbf{x}$ under σ lies in the orbit of $\mathbf{x}$ under G; and (G, X) is k-closed if it equals its k-closure. This is an important notion for us, as it recaptures some information about the original presentation of the structure which one might have expected to see washed away by the Galois connection.

This may be refined as follows: the *complexity* of a permutation group (G, X) is the pair (k, r), where k is minimal such that G is k-closed, and r is minimal number such that there are r $(\leq k)$-ary invariant relations in a structure $\mathcal{X}$ for which Aut $\mathcal{X} = G$. The signature of a structure and the complexity of the associated permutation group are closely related; the complexity of a permutation group measures the size of the simplest signature σ such that some structure of signature σ has the specified group as its automorphism group. In model theory, the signature is taken as given at the outset; in permutation group theoretic terms, this amounts to bounding the complexity.

Another invariant of k-closed groups which is relevant here is the number r_k of orbits of G on X^k; these orbits are called k-*types* and may defined in structural terms as the atoms in the boolean algebra of invariant (= definable) relations. We will have $r \leq r_k$ and in the context of homogeneous structures, $r_k \leq 2^{k2^r}$, so once k is bounded, bounding the complexity or bounding r_k amounts to the same thing.

A finite relational structure is *homogeneous* if every isomorphism between induced substructures is the restriction of an automorphism. We say that a finite relational structure is k-*ary* if it is equivalent (or, with some abuse of language: isomorphic) to a homogeneous structure whose relations are k-ary. On the permutation group theoretic side, we may translate this to the following condition which is rather tricky to analyze in practice: (G, X) is k-ary if for any $n \leq |X|$, if $\mathbf{a}$, $\mathbf{b}$ are two n-tuples with the property that all pairs of corresponding subsequences of $\mathbf{a}$, $\mathbf{b}$ of length at most k lie in the same G-orbit, then $\mathbf{a}$ and $\mathbf{b}$ lie in the same G-orbit. If a group is k-ary then it is k-closed, but the converse is thoroughly false. The automorphism groups of the graphs n^2 in the preceding section are 2-closed by definition, but the main point made about them there was that they are usually *not* binary in our sense (in other words, they are not homogeneous as binary structures); they are 4-ary structures.

Aschbacher refers to homogeneity as the *Witt property* in the passage cited in our epigraph, which continues: "If X is an object with the Witt property and G is its group of automorphisms, then the representation of G on X is usually an excellent tool for studying G." Indeed.

The notions of homogeneity and k-arity are the starting point for Lachlan's theory, which concerns the class of k-ary structures for k fixed, and with a bound on r_k also fixed.

Notation Let $\mathcal{X}$ be a structure. We write $\kappa(\mathcal{X})$ for the *degree of homogeneity* of $\mathcal{X}$, which is the least k for which $\mathcal{X}$ is k-ary.

While Lachlan's theory provides a good classification of structures with $\kappa(\mathcal{X})$ and $r_{\kappa(\mathcal{X})}$ bounded, and while every finite structure comes into this classification eventually, with $\kappa(X) \leq |\mathcal{X}|$, the theory does not provide any information directly about the point at which a given structure of mathematical interest will occur, that is: how is $\kappa(\mathcal{X})$ computed for interesting structures (or permutation groups) $\mathcal{X}$?

We have now introduced our basic vocabulary, but as we noted in passing above, something odd happens with the notion of isomorphism, and this is worth dwelling on. Two permutation groups (G, X) and (H, Y) will be considered isomorphic if there is a bijection of X with Y carrying G to H. This notion will be carried over to structures: $\mathcal{X}$ will considered isomorphic to $\mathcal{Y}$ (via the map $f : X \leftrightarrow Y$) if (Aut $\mathcal{X}, X$) is isomorphic to (Aut $\mathcal{Y}, Y$) (via the same map). For example: a graph is isomorphic to its complement via the identity map; the graphs n^2 are isomorphic to their enrichments by the parallelism relation; a finite set carrying a successor relation is isomorphic to the same set equipped with the induced linear order. Graph theorists may well disagree with the first of these examples, but from our point of view there are two nontrivial relations between vertices, and it is not of much importance which one is *called* the edge relation; in the canonical structure, this amounts to permuting the names by which various relations are known. This cavalier attitude is appropriate in dealing with homogeneity, and more generally with any issues that can be understood at the level of permutation groups. Once one is committed to this notion of isomorphism, one tends to replace the terms "relational structure" and "permutation group" by "permutation structure", and to lose track of which side of the Galois connection one is actually working with at any given moment.

In §1 we encountered an example of the *wreath product* construction. On the permutation group side, one begins in general with two permutation groups (G, X) and (H, I), called, respectively, the *base* group and the *index* group. One then forms the *wreath product* $(G \wr H, X^I)$ which is set-theoretically a power on which G^I acts coordinatewise, while H permutes the coordinates. Thus $G \wr H = G^I \rtimes H$. The same construction applied to a base structure $\mathcal{X}$ and an index structure $\mathcal{I}$ yields a canonical structure $\mathcal{X}^{\mathcal{I}}$, which may be replaced in practice by some other structure with the same automorphism group.

There are two extreme cases: let d represent the permutation group $\text{Sym}(d)$ acting naturally on $\{1, \ldots, d\}$ (as a structure this is a bare set, carrying only the equality relation), and let $\bar{d}$ be the trivial group acting on the same set, which in structural terms is a labeled set whose elements are treated as distinguished constants. The power $\mathcal{X}^d$ is the traditional power which can be given by lifting each of the relations on $\mathcal{X}$ (including the equality relation) to d relations operating in the d possible "directions". In particular the equality relation lifts to n equivalence relations. The power $\mathcal{X}^{\bar{d}}$ is the symmetrized power in which the coordinates may be freely permuted. In particular n^d was presented concretely in §1: it is the wreath product of two sets with no additional structure. These are among the simplest structures occurring in nature, though by no means the simplest structures from the point of view of the computation of $\kappa(\mathcal{X})$.

Some useful permutation group theoretic notions have no apparent analog on the structural side of the Galois connection: notably the *socle*, and more generally the notion of a *normal subgroup*. In particular the structures corresponding to the natural representations of $\text{Sym}(n)$ and $\text{Alt}(n)$ on $\{1, \ldots, n\}$ have virtually nothing in common from the point of view that interests us: the former is degenerate, and the latter has κ maximal. On the other hand, as we will see, there are excellent reasons for returning to the structural side of the picture, particularly as we exploit the possibility of taking infinite limits of our structures; though the Galois connection survives in the infinite limit, with some modifications, still its group theoretic side becomes largely useless. To make the connection work well with X infinite, one takes into account the natural topology on $\text{Sym}(X)$, and one restricts attention to closed groups having finitely many orbits on X^k for each k. For some of this, see [DM] or [Ca2]. In any case we will not actually exploit this connection in infinite structures.

There is one quite reasonable operation which can behave atrociously, regardless of which side one operates on: formation of quotients. This is of some technical importance, and one of the main theorems in the subject is concerned with the structure of such quotients (Proposition 1, §3). In a permutation structure $\mathcal{X}$, the invariant equivalence relations form a lattice with 0 and 1; the structure is primitive if this is the whole lattice. If E_1 and E_2 are two invariant equivalence relations with $E_1 \geq E_2$ in this lattice, and if C is an E_1-class, then the quotient C/E_2 is naturally a permutation structure: the group acting on C/E_2 is the faithful version of the group induced on C/E_2 by the setwise stabilizer in $\text{Aut}\,\mathcal{X}$ of C. When E_1 covers E_2 and E_2 is nontrivial on C, this quotient is primitive. Some properties are inherited, notably bounds on the number of k-types for each k. The properties of k-closure and k-arity behave as badly as one could imagine.

In fact:

Every finite structure is a quotient of a finite binary structure.

For example if the automorphism group is transitive, and (G, X) is the corresponding permutation group, then the structure is naturally a quotient of the right regular representation of G, which is binary homogeneous (equip the underlying set of G with one binary relation for each element g of G, which encodes the action of g by right multiplication).

Another complication: the Jordan-Holder theorem fails badly in this context. Consider for example the structure $\mathcal{X}_n$ consisting of ordered pairs from $\{1, \ldots, n\}$ with distinct entries, on which $\mathrm{Sym}(n)$ acts naturally. If E is a nontrivial invariant equivalence relation on $\mathcal{X}_n$, one can analyze the structure as an extension of the quotient $\mathcal{X}_n/E$ by the structure on an E-class. If E is the relation of having an identical first coordinate, this means $\mathcal{X}_n$ is treated as an extension of n by n. If we consider instead the relation E' of corresponding to the same unordered pair, then $\mathcal{X}_n$ becomes a 2-fold cover of n^2, which is a primitive structure; furthermore $\mathcal{X}_n$ is binary homogeneous, and in just one of these two analyses the components are also binary homogeneous. In spite of this, one can still get useful information using induction on the length of such (non-unique) composition series.

The next section contains a variety of examples illustrating the behavior of $\kappa(\mathcal{X})$. This is a digression as far as the general theory is concerned; we return to the main line in §4 with the classification of the finite homogeneous graphs, in which the outlines of a general theory of finite homogeneous structures are very dimly visible – sufficiently visible to Lachlan, in any case, to spark the development of that theory.

3 $\kappa(\mathcal{X})$: examples

Let us write $k(\mathcal{X})$ for the least k such that $\mathrm{Aut}(\mathcal{X})$ is k-closed, and $\kappa(\mathcal{X})$, as in the previous section, for the least κ such that $\mathcal{X}$ has a presentation as a κ-ary structure; equivalently, this means that every $\mathrm{Aut}(\mathcal{X})$-invariant relation is a boolean combination of $\mathrm{Aut}(\mathcal{X})$-invariant κ-place relations. The present section will offer a smørgasbord of examples illustrating the behavior of $\kappa(\mathcal{X})$, and the reader is invited to consult his own appetite.

The computation of both $k(\mathcal{X})$ and $\kappa(\mathcal{X})$ present substantial difficulties, but the basic meaning of the invariant $k(\mathcal{X})$ seems the more accessible. For example, for $k = 2$: a permutation group is 2-closed if it is the automorphism group of the directed graph with colored edges (some symmetric, some asymmetric) obtained by taking the orbits of the group on ordered pairs, and giving each orbit its own color. On the other hand, a permutation group G is *binary* if every G-invariant relation is a boolean combination of G-invariant binary relations. In practice, this gets decoded as follows: (G, X) is *not* binary if:

there are two ordered sequences $a_1, \ldots, a_r$ and $b_1, \ldots, b_r$ of points of X of length $r > 2$, not conjugate under the action of G, such that:

any ordered pair of elements from **a** is conjugate to the corresponding pair from **b** under the action of G.

Note that if $r > 2$ is minimal with this property, then we will have a stronger condition: any sequence of $r - 1$ elements of **a** is conjugate under G to the corresponding subsequence of **b**. In this case we might as well take $a_i = b_i$ for $i \leq r-1$ here. So we may change to the following notation: $a_1, \ldots, a_{r-1}, b$ and $a_1, \ldots, a_{r-1}, b'$ are the two sequences, and our condition becomes (for some $r > 2$):

b and b' lie in distinct orbits over $a_1, \ldots, a_{r-1}$,

but in the same orbit over any $r - 2$ of the elements a_i (∗)

where the orbit "over" a set of points is the orbit under the pointwise stabilizer of that set in G. An advantage of the last formulation is that with r fixed, it expresses: $\kappa(\mathcal{X}) \geq r$; so it is no longer tied to the case $\kappa = 2$.

Example 1 If $\mathcal{X}$ is a naked set (i.e., $\mathrm{Aut}(\mathcal{X}) = \mathrm{Sym}(X)$), then $\kappa(\mathcal{X}) = 2$.

This is intuitively obvious and can be read off of (∗) directly; the orbit of b over $a_1, \ldots, a_{r-1}$ is determined by its orbit over each of the a_i (which amounts to determining whether b is equal to one of the a_i).

Example 2 $\kappa(\mathrm{Alt}(n), X) = n - 1$ where $\mathrm{Alt}(n)$ acts naturally on $X = \{1, \ldots, n\}$.

Just look at the sequences $(1, \ldots, n - 2, n - 1)$ and $(1, \ldots, n - 2, n)$. These demonstrate that $\kappa(\mathrm{Alt}(n), X) \geq n - 1$. The reverse inequality is equally evident, by the same test. (And $\kappa(G, X) < |X|$ for any permutation group, for the same reason.)

Example 3 Let $\mathcal{V}$ be $(\mathrm{GL}(V), V)$ with the natural action.   Then $\kappa(\mathcal{V}) = \dim V + 1$ if the base field has more than 2 elements, and is $\dim V$ otherwise.

The relevant pair of sequences $\mathbf{a}, b$; $\mathbf{a}, b'$ is gotten by taking $\mathbf{a}$ to be a basis, $b = \sum_i a_i$, and b' some other linear combination with all coefficients nonzero, assuming the base field has more than two elements. Otherwise one takes $\mathbf{a}$ to be a basis with one element removed; $b = \sum_i a_i$; and b' is an additional basis element. This provides the relevant lower bound for κ in each case, and the upper bound is a triviality; in fact the upper bound $\dim V + 1$ will hold for any group of linear transformations on V (though not necessarily for a group of semilinear transformations).

This is a good example, because $\kappa(\mathcal{X})$ is a measure of complexity which is very like the dimension of an arbitrary finite structure, but taking into account all invariant relations among elements, and not just those provided in the original description of the structure $\mathcal{X}$. The next example continues this line of thought.

Example 4   [CMS] Let $\begin{bmatrix} n \\ k \end{bmatrix}$ be the permutation group $\mathrm{Sym}(n)$ acting on the set of k-sets in a set $X = \{1, \ldots, n\}$, with $n \geq 2k$ (or $n > 2k$ if one insists on primitivity). Then $\kappa(\begin{bmatrix} n \\ k \end{bmatrix})$ is $\lceil \log_2 k \rceil + 2$.

We may think of $\begin{bmatrix} n \\ k \end{bmatrix}$ also as a graph: two k-sets may be taken to be adjacent if they are disjoint (e.g., for $n = 5$, $k = 2$: the Petersen graph). This is the same structure (has the same group of automorphisms), so the group is 2-closed. According to the formula, though, $\kappa(\begin{bmatrix} n \\ k \end{bmatrix}) > 2$ for $k > 1$. For example, $k = 2$: $\kappa(\begin{bmatrix} n \\ k \end{bmatrix}) = 3$ in this case, and this is illustrated by the triples $\{1,2\}; \{1,3\}; \{1,4\}$ and $\{1,2\}; \{1,3\}; \{2,3\}$. It requires a little more care to check that $\kappa \leq 3$ in this case.

For the general case, one first gives a similar example showing $\kappa(\begin{bmatrix} n \\ k \end{bmatrix}) \geq \lceil \log_2 k \rceil + 2$. The work comes in the reverse inequality. One notes that the orbit of a sequence of k-sets is determined by the cardinalities of the atoms in the boolean algebra they generate (we allow degenerate atoms which are empty; they are really labelled by atoms in the free boolean algebra on the same number of generators). One has to show that these numbers are determined by the corresponding data for the subalgebras generated by $\lceil \log_2 k \rceil + 1$ elements; the main point is just that whenever a set is split into two pieces, one of the pieces is at most half as large as the original.

These particular structures play a fundamental role in the general theory, where they occur as "grassmannian" structures (motivated terminologically by the fact that they are the $q = 1$ versions of grassmannians in vector spaces over $\mathbb{F}_q$).

Example 5   Let $\begin{bmatrix} n \\ k \end{bmatrix}^\circ$ be the permutation group $\mathrm{Alt}(n)$ acting on the set of k-sets in a set $X = \{1, \ldots, n\}$, with $n \geq 2k$. Then $\kappa(\begin{bmatrix} n \\ k \end{bmatrix}^\circ) =$

$\max(n - k, n - 3)$, except for $k = 2$, $n = 4$, where the value is 3.

One gets lower bounds for κ from examples, and fortunately the values are high enough that one can get matching upper bounds without getting too badly bogged down. We will give some details, since this has not been documented elsewhere.

One may assume $k \geq 2$, and the case $k = 3$, $n = 6$ is best inspected separately.

The lower bounds

For $k = 2$ it suffices to consider the induced action on $\{\{i, n\} : i < n\}$, which is equivalent to $\mathrm{Alt}(n - 1)$ in its natural representation. This gives the lower bound $\kappa \geq n - 2$ in this case, by example 2. For $n = 4$ one examines the orbits of a specific pair of sequences: $\{1, 2\}$, $\{1, 3\}$, $\{1, 4\}$ and $\{1, 2\}$, $\{1, 3\}$, $\{2, 3\}$; the same ones which would be used for the full symmetric group.

For $k \geq 3$ we need an explicit example. We set $a_i = \{i\} \cup \{k, k + 1, \ldots, 2k - 2\}$ for $i \leq k - 2$, and $a_i = \{1, \ldots, k - 1\} \cup \{i + 2\}$ for $k - 1 \leq i \leq n - 4$. Taking $b = \{1, \ldots, k - 2\} \cup \{n - 1\}$ and $b' = \{1, \ldots, k - 2\} \cup \{n\}$, we claim that b and b' lie in distinct orbits over $\mathbf{a} = (a_1, \ldots, a_{n-4})$ and in the same orbit over any proper subsequence. To see this one has to compute the pointwise stabilizer in $\mathrm{Alt}(n)$ of $\mathbf{a}$ and of its subsequences; this amounts to computing the boolean algebra generated by one of these subsequences, or at least getting sufficient information about the atoms. The atoms of the algebra generated by $\mathbf{a}$ are $\{i\}$ for $i \leq n - 2$ together with $\{n - 1, n\}$; the pointwise stabilizer of this algebra in $\mathrm{Alt}(n)$ is trivial. Over proper subsequences one identifies larger atoms easily, and one then sees that b, b' lie in the same orbit for their stabilizers in $\mathrm{Alt}(n)$. The case $k = 3$, $n = 6$ should be treated separately from the general case.

This analysis shows that the value we have given for $\kappa(\begin{bmatrix} n \\ k \end{bmatrix}^\circ)$ is a valid lower bound.

The upper bounds

To get matching upper bounds is more troublesome. One considers any pair of sequences $a_1, \ldots, a_\kappa$, $b_1, \ldots, b_\kappa$ illustrating that $\kappa \geq \kappa(\begin{bmatrix} n \\ k \end{bmatrix}^\circ)$, and one shows that $\kappa \leq n - 2$, and $\kappa \leq n - 3$ for $k \geq 3$. Leaving aside the case $k = 3$, $n = 6$, this is done by looking closely at the sequence $a_1, \ldots, a_\kappa$, which has the following property, without loss of generality:

a_i is not in the boolean algebra generated by $(a_j : j \neq i)$ for any i (∗)

Indeed, if this fails then a_i is fixed over $(a_j : j \neq i)$ by the full symmetric group, and then Example 4 applies to show $\kappa \leq \lceil \log_2 k \rceil + 2$, a bound which yields $\kappa \leq n - 2$ for $k = 2$, apart from the known exception $n = 4$;

and the same estimate yields $\kappa \le n - 3$ for $k \ge 3$ and $n \ge 2k$. So we need only consider the case $(*)$.

As $k \ge 2$, the a_i are not all disjoint, so we may suppose a_1 meets a_2; then the boolean algebra generated by a_1, a_2 will have 4 atoms, and applying $(*)$ the boolean algebra generated by the a_i will have at least $\kappa + 2$ atoms, proving $\kappa \le n - 2$ and indicating that the case $\kappa = n - 2$ is extreme.

Suppose now that $k \ge 3$. We must eliminate the possibility $\kappa = n - 2$. One finds a_1, a_2, a_3, a_4 generating 7 atoms, which suffices; in fact one finds a_1, a_2, a_3 generating 6 atoms in many cases. A useful observation is that the a_i separate points as there are $n - 2$ of them and the algebra they generate must have at least n atoms.

One should dispose first of the case $n = 2k$, $k \ge 4$, leaving the case $n = 6$, $k = 3$ to be dealt with by inspection. The methods are much the same as those used in the main case: $k \ge 3$, $n > 2k$.

Suppose first that one can choose a_1, a_2 so that $|a_1 \cap a_2| = k - 1$. As the a_i separate points, we may take a_3 splitting $a_1 \cap a_2$. Then a_i cannot split $(a_1 \cup a_2)'$ as otherwise we already have 6 atoms. On the other hand $|(a_1 \cup a_2)'| \ge k$ so a_i cannot contain $(a_1 \cup a_2)'$ either, and $a_3 \subseteq (a_1 \cup a_2)$. Hence a_i consists of $k - 2$ elements of $a_1 \cap a_2$ and the two elements of $(a_1 \cup a_2) \setminus a_1 \cap a_2$. In particular this analysis shows that $a_1 \cap a_2$ is not contained in any other a_j, as otherwise the same analysis would apply with a_j in place of a_2.

Now one uses fully the fact that the a_i separate points. Let $A = a_1 \cup a_2$. We may suppose that $A = \{1, \ldots, k + 1\}$, and then after relabeling that $a_i = A - \{i\}$ for $i \le k$. Then if a_i splits A', we find $a_i \cap A = \emptyset$ as otherwise we can choose $i_1 < i_2 \le k$ so that a_{i_1}, a_{i_2}, a_i generate at least 6 atoms. If now follows just by counting that $\kappa \le n - 3$; in addition to the k elements a_i already determined above, there are at most $n - k - 3$ disjoint from A, using $(*)$.

The rest goes similarly and more quickly. Taking a_1, a_2 so that $l = |a_1 \cap a_2| > 0$ is minimized, one finds easily that $l = 1$ (distinguishing the cases $l > k/2$, $l \le k/2$ along the way). Then splitting $(a_1 \cup a_2)'$ and $a_1 \setminus a_2$ by elements a_3, a_4, one either gets the desired subalgebra on 3 or 4 generators, or in the remaining case one finds $|a_3 \cap a_4| = k - 1$, the case treated at the outset. $\square$

We now consider the binary case further. An *affine* group of dimension d is a subgroup of $A\Gamma L(V)$ containing the translation subgroup V, with V d-dimensional; it is *strictly linear* if it is a subgroup of $AGL(V)$. Here $AGL(V) = V \rtimes GL(V)$ and $A\Gamma L(V) = V \rtimes \Gamma L(V)$.

Example 6 A primitive 1-dimensional strictly linear affine group G is binary if and only if it is cyclic or dihedral; otherwise $\kappa(G) = 3$.

The structures in the cyclic or dihedral cases are directed or undirected cycles, respectively, of prime order.

Note that the stabilizer of any two points is trivial in this case. This already forces $\kappa(G) \leq 3$. So the only point is to identify the binary cases. Since the group is 1-dimensional, we will denote it $\mathbb{F} \rtimes \mu$ where $\mathbb{F}$ is the additive group of the base field and μ is a subgroup of the multiplicative group. We assume $|\mu| > 2$ and we show that G is not binary. Consider the triples $(0, -1, g)$ and $(0, -1, g^{-1})$ with $g \in \mu$, $g \neq \pm 1$. Since the stabilizer of two points is trivial and $g \neq g^{-1}$, these triples lie in distinct orbits. However the pairs $(0, g)$ and $(0, g^{-1})$ lie in the same orbit under μ, and the pairs $(-1, g)$ and $(-1, g^{-1})$ lie in the same orbit under G – translate by $+1$ and multiply by g^{-1}. Thus we have an explicit violation of binarity.

We will push this a bit further because it will complete the list of currently known primitive binary permutation groups; and the determination of all such would be very welcome.

Example 7 Let G be a primitive 1-dimensional affine group, not strictly linear. Then $\kappa(G) \leq 4$ and G is binary if and only if G has the form $\mathbb{F}_{q^2} \rtimes \mu_{q+1} \rtimes \langle \sigma \rangle$ with σ of order 2.

Proof. For the upper bound, one shows that the permutation group given by the stabilizer of two points is binary. (Any action of a cyclic group is binary.) The typical value is probably 4 and one may be able to identify all the exceptions, but this has not been done. For our purposes, it suffices to identify the binary ones. We leave to the reader the computation that the the groups listed are binary; we will analyse these examples a little more below. The main point is that no other binary examples are to be found in this class.

Suppose G is binary, $G = \mathbb{F} \rtimes H$ with $H \leq \mathbb{F}^{\times} \rtimes \Gamma$, $\Gamma = \text{Aut } \mathbb{F}$. Then the argument given in the linear case applies, but shows only that for $h = a\sigma \in H$, $1^h = 1^{h^{-1}}$, i.e. $a^{\sigma} = a^{-1}$. Let μ be the projection of H on $\mathbb{F}^{\times}$ and let Γ_o be the projection of H on Γ; then for $\sigma \in \Gamma_o$, we have just seen that σ^2 fixes μ. However the field generated by μ^{Γ_o} is $\mathbb{F}$, by primitivity, so $\sigma^2 = 1$. Thus $|\Gamma_o| \leq 2$ and, since G is not strictly linear, $\Gamma_o = \langle \sigma \rangle$ with σ of order 2. Thus $\mathbb{F} = \mathbb{F}_{q^2}$ for some prime power q, and $a^{\sigma} = a^{-1}$ for $a \in \mu$.

Thus H is a subgroup of the desired group $H_1 = \mu_{q+1} \rtimes \langle \sigma \rangle$. Take $a\sigma \in H$. Then $a^{\sigma} = a^{-1}$ so $a = b/b^{\sigma}$, some b, and conjugating G by b (as an element of $\mathbb{F}^{\times}$), we may take $\sigma \in H$. Thus $H = \mu_o \rtimes \langle \sigma \rangle$ with $\mu_o \leq \mu_{q+1}$. We will show $\mu_o = \mu_{q+1}$.

As G is primitive, μ_o contains some $r \neq \pm 1$. Let $s \in \mu_{q+1}$ be arbitrary. Solve $b^\sigma/b = s$ for b. Consider the triples $(0, b, b/(r+1))$ and $(0, bs, bs/(r+1))$. Any pair from the first triple is carried to the corresponding pair of the second triple by one of the maps σ, σr, or σr^{-1}. By binarity there is a transformation $g \in G$ taking $(0, b, b/(r+1))$ to $(0, bs, bs/(r+1))$. Then $g \in H$, and since $b^g = bs$ it follows that $g = \sigma$ or $g = s$. As $[b/(r+1)]^g = bs/(r+1)$ we can exclude the former possibility and thus $g = s$. So $s \in G$, for any such s. $\qquad\square$

Remark Let Γ_q be the binary structure corresponding to the binary 1-dimensional affine group $\mathbb{F}_{q^2} \rtimes \mu_{q+1} \rtimes \langle\sigma\rangle$. Then Γ_q is a symmetric graph with an edge coloring by $q - 1$ colors.

The symmetry means that we can solve the equation $a^{\sigma r} = -a$ with $r \in \mu_{q+1}$ for any a. This just means $-a^\sigma/a \in \mu_{q+1}$, which is the case. This also shows that the orbit over 0 of any point has order $q+1$ and thus there are $q - 1$ such orbits.

In particular we have an ordinary uncolored graph only for $q \leq 3$. The case $q = 2$ is degenerate and for $q = 3$ the associated graph is the sporadic homogeneous graph K_3^2. This is a second way of accounting for this example, quite different from that of §1. In the case $q = 4$ one has three colors, and again this example turns up as one of a small number of sporadics in the classification corresponding to that case, which was carried out long ago by Lachlan. This particular example is also used to show that the Ramsey number $r(3, 3, 3) \geq 17$, as it provides a graph of order 16 with a 3-edge coloring without chromatic triangles (an example given by Andrew Gleason).

We do not know of any other finite primitive binary homogeneous structures, apart from those we have seen: cyclic, dihedral, Γ_q, and of course $\text{Sym}(n)$ acting naturally.

Example 8: n^d Our final example is considerably more subtle. Recall that n^d is the wreath product (acting on a power) of the natural representations of $\text{Sym}(n)$ and $\text{Sym}(d)$. The value of $\kappa(n^d)$ has been worked out by Saracino [Sa]. [CMS] contains estimates of arities of wreath products $\mathcal{X}^{\mathcal{Y}}$ in general, and shows that the upper bound given there provides the exact value for "most" values of n and d, specifically: for $n \geq 2[\log_2 d]+2$.

The upper bound is given in general by

$$\kappa(\mathcal{X}^{\mathcal{Y}}) \leq \kappa(\mathcal{X}) \cdot \kappa(\mathcal{P}(\mathcal{Y}))$$

where $\mathcal{P}(\mathcal{Y})$ is the power set of $\mathcal{Y}$ (with automorphism group $\text{Aut}(\mathcal{Y})$, acting naturally). We have $\kappa(n) = 2$ and $\kappa(\mathcal{P}(\{1, \ldots, d\})) = [\log_2 d] + 1$

[CMS], and thus $\kappa(n^d) \leq 2(\lceil \log_2 d \rceil + 1)$; and this turns out to be the exact value for $n \geq 2\lceil \log_2 d \rceil + 2$.

The value of $\kappa(n^d)$ for relatively small values of n follows a more complicated rule, which takes on a distinctly simpler form if we look instead for a formula which computes, in terms of given n and κ, the minimum value of d for which $\kappa(n^d) \geq \kappa$. For technical reasons it is better to define a very similar function $\delta(\kappa, n)$ as the least value of d such that there are two sequences of length κ in n^d of length κ which witness that $\kappa(n^d) \geq \kappa$, in the sense of $(*)$ above, and which are not conjugate to sequences occurring in $(n-1)^d$. The most important point here is that we choose to express d in terms of n and κ, getting a moderately complex set of conditions which can be explicitly but more awkwardly solved for κ in terms of n and d.

The main formulas that result are:

$$\delta(\kappa, n) = \begin{cases} 2^{\kappa - n/2 - 1} & \text{for } \kappa > n \text{ even} \\ 3 \cdot 2^{r - \frac{n+5}{2}} & \text{for } n \geq 5 \text{ odd and } \kappa > n \text{ even} \\ 9 \cdot 2^{r - \frac{n+7}{2}} & \text{for } \kappa > n \geq 5 \text{ with } \kappa \text{ and } n \text{ odd} \\ 27 \cdot 2^{r - n/2 - 5} & \text{for } \kappa > n \geq 8 \text{ with } \kappa \text{ odd and } n \text{ even} \end{cases}$$

with similar formulas, and some exceptional values, covering the remaining cases.

The method used is to attach a clearcut combinatorial invariant to the orbit of an r-tuple in a wreath product. This can be done quite generally, though it is rather messy in general. In the case of n^d the type of an r-tuple can be encoded by a multiset consisting whose elements are equivalence relations on the set $\{1, \ldots, r\}$ having at most n classes (and the final, technical part of the definition of δ corresponds to the condition that at least one of these relations should actually have n classes). The data referred to in condition $(*)$ above would then be two such multisets which coincide whenever a single index i (with $1 \leq i \leq r$) is deleted from all the relations in both multisets). The parameter δ which is sought is the number of relations occurring in each of these multisets; examples show that the stated values are valid upper bounds, and one must then prove matching lower bounds for the sizes of such multisets. This is what Saracino has done.

All of this yields:

For $n \leq 2\lceil \log_2 d \rceil + 2$,
$$\kappa(n^d) = 2\lceil \log_4 \alpha_n 2^{n/2+1} d \rceil + \epsilon$$
with $\epsilon = 0$ or 1 unless $n = d = 3$,
and with $\alpha_n = 1$ for n even and $4/3\sqrt{2}$ for n odd.

For $n \geq 2\lceil \log_2 d \rceil + 2$,

$$\kappa(n^d) = 2\lceil \log_2 d \rceil + 2$$

Thus a bound on κ, as specified at the outset in one of Lachlan's classification problems, will pick out: (1) the families n^d for which n is arbitrary and $d < 2^{\kappa/2}$; and (2) the structures n^d for which $n \leq 2\lceil \log_2 d \rceil + 2$ and $2\lceil \log_4 \alpha_n 2^{n/2+1} d \rceil + \epsilon \leq \kappa$, i.e. roughly speaking $n/2 + 1 + \lceil \log_2 d \rceil \leq \kappa$. E.g. for $\kappa \leq 10$, the infinite families are n^d for $d \leq 31$, and the last sporadic structure would be 2^{511}; so here the number of sporadics is exponentially greater than the number of well-behaved families. (Note that a bound on r_κ is not of great importance in this particular context.)

Other examples which one would naturally consider include multiply transitive groups, and the action of $\mathrm{Sym}(n)$ or $\mathrm{Alt}(n)$ on partitions of $\{1, \ldots, n\}$ of a fixed type. The multiply transitive representations do not generally present any particular difficulties: the sharply transitive ones of degree t have $\kappa = t + 1$, apart from the natural representation of $\mathrm{Sym}(n)$, and in non-sharply transitive cases the value is at least $t + 1$, and not much greater. Actions on partitions are not well understood at all, and to round out this collection one would like to have decent estimates for this case (at least under the full symmetric group) and for corresponding wreath products. A few such problems are discussed further in §9.

4 Finite and smoothly approximable homogeneous graphs

We will now present the classification of the finite and countably infinite homogeneous graphs, beginning with the finite ones. It will then be obvious that the list of finite homogeneous graphs can be completed by adjoining their natural infinite limits, and we will need to give a precise definition which describes this completion process in a general setting, in terms of a notion of *smooth approximability*. For homogeneous structures this property is equivalent to one of the fundamental notions of pure model theory: *stability*. There are other infinite homogeneous graphs that fall outside the smoothly approximable scheme; though these have also been classified explicitly, this part of the classification has not been subsumed by any more general theory.

The finite homogeneous graphs

1. $m \cdot K_n$: *the disjoint sum of m complete graphs, each of order n.*

1'. $(m \cdot K_n)'$: *the complementary graphs, complete m-partite graphs with parts of order n.*

2. *The pentagon C_5.*

3. $3^2 = K_3^2 = L(K_{3,3}) = \Gamma_3$. *This has been described at various points above as a wreath product of degenerate structures, the line graph of a complete bipartite graph $K_{3,3}$, or an unusual binary structure associated with a 1-dimensional affine but not strictly linear group over $\mathbb{F}_3$.*

Actually in the language of §2, the graphs of types 1 and 1' are isomorphic structures; we view antiisomorphisms as a permutation of the language. But this still does not entitle us to view them as isomorphic *graphs*, so one often carries the complementary family along. Each of the sporadic examples (2) and (3) is self-dual. These two graphs can be viewed as degenerate members of families occurring naturally with $\kappa > 2$, in a way which is interesting in the case of 3^2 but much less so for C_5, where the natural family to consider would be one in which the points of C_5 are replaced by equivalence classes of arbitrary size. Viewing C_5 as part of the family $\{C_n\}$ and 3^2 as part of the family $\{\Gamma_q\}$ is also very reasonable, but this does not correspond to the hierarchy by degree of homogeneity in the form it to which Lachlan's finiteness theorems apply: in the families $\{C_n\}$ and $\{\Gamma_q\}$, we have $\kappa = 2$, but r_κ is unbounded.

One rounds out this list by allowing m and n to become countably infinite in families 1 and 1'. (However we require them to be countable.) This process needs to be described more intrinsically, and the following definition expresses what we will mean by the *infinite version* of a family of finite structures, in general.

Definition   Let $\mathcal{X}$ be a structure such that $r_k(\mathcal{X})$ is finite for all k.

1 The structure $\mathcal{X}$ is said to be a *smooth limit* of finite structures if every finite subset $X_\circ$ of the universe X is contained in a set X_1 for which the induced structure $\mathcal{X}_1$ is "smoothly embedded" in $\mathcal{X}$, where the latter condition is defined as follows.

2 $\mathcal{X}_1$ is *smoothly embedded* in $\mathcal{X}$ if any two finite sequences of elements of $\mathcal{X}_1$ which lie in the same orbit under $\mathrm{Aut}(\mathcal{X})$ also lie in the same orbit under $\mathrm{Aut}(\mathcal{X}_1)$, where $\mathrm{Aut}(\mathcal{X}_1)$ is the group induced on the underlying set X_1 by its setwise stabilizer in $\mathrm{Aut}(\mathcal{X})$. ($\mathcal{X}_1$ is an induced substructure of $\mathcal{X}$).

The finiteness assumption on all r_k is harmless in the context of homogeneous structures with κ and r_κ bounded; any smooth limit of such structures will inherit the same bound on κ and r_κ, and homogeneity then implies that r_k is bounded for all k. Smooth approximability is a key condition which is of interest outside the homogeneous context as well; other examples are provided by infinite-dimensional vector spaces over finite fields, which may be decorated with the inner products or quadratic forms which define the various classical groups.

The consideration of these infinite analogs of the finite structures is quite helpful. For example we can reduce list (1) above to the finite smoothly embedded substructures of one structure: $\infty \cdot K_\infty$. It can be convenient to replace infinite families of finite structures by a single infinite limit.

We have mentioned that there are other countably infinite homogeneous infinite graphs, and that these have also been classified. One such is Rado's graph, or "the" random graph, which may be described as follows: if a graph G is constructed by putting in edges randomly and independently, with constant probability p ($0 < p < 1$), then there is a single graph G_∞ such that with probability 1 the random graph G is isomorphic to G_∞. For each n there is a similar graph G_n, called the generic countable graph, which contains no clique of order n; this cannot be defined probabilistically, but can be defined using topology in place of measure theory. One views the collection $\mathcal{G}_n$ of K_n-free graphs as a compact topological space, and one looks for a graph G_n such that the set of graphs in $\mathcal{G}_n$ which are not isomorphic to G_n is topologically meager. There is a better description of these graphs in terms of amalgamation classes (§6) but this will certainly do for the moment.

The full classification of the homogenous graphs is then:

I. The homogeneous graphs which are smoothly approximated by finite graphs;

II. The generic graph omitting the n-clique K_n, for fixed $n \geq 3$;

II′. The complements of the graphs of type II;

III. The Rado graph (self-dual).

If the Rado graph is an unfamiliar object, one can approach it by considering the rational order $(\mathbb{Q}, <)$ as an analog. Any ordered set can be viewed as a directed graph (in fact a tournament) and one can consider the homogeneous orders. It follows directly from the definition that the only homogeneous orders are: (1) the "order" on one point; and (2) a dense linear order without endpoints, which may be taken to be $(\mathbb{Q}, <)$. The dichotomy occurring in the theory of homogeneous graphs occurs here in an extreme form, as there is one extremely finite example and one extremely infinite example; $(\mathbb{Q}, <)$ has no finite smooth approximation with more than one element. Just as the ordering of $\mathbb{Q}$ can be characterized by its density properties, the Rado graph can be characterized by analogous density properties, stating that any finite subgraph can be extended in all possible ways by the addition of a suitable additional vertex. Peter Winkler and other fans of Arlo Guthrie call this the Alice's restaurant property.

The part of this classification which goes beyond the smoothly approximable case is found in [LW]. Some other classification results of a similar character have been found; the proofs are purely combinatorial, relying on Ramsey's theorem, and are usually long. One may also detect in the case at hand a striking dichotomy, an instance of a more systematic dichotomy in model theory uncovered by Shelah: the stable/unstable distinction. This more technical idea enters heavily into the proofs, and occasionally into the statements, of the main results.

One approach to stability is to define one or more notions of dimension for arbitrary structures, referred to in model theory as ranks – a confusing terminology when used in connection with permutation structures – and to call a structure stable if the rank or ranks used are finite. Following Lachlan, we will use a single notion of rank which is well adapted to the group theoretic viewpoint. We will call this particular notion the *orbit height*.

Definition Let $\mathcal{X}$ be a homogeneous structure with $\kappa(\mathcal{X})$ and $r_{\kappa(\mathcal{X})}$ finite.

1. A *tree of orbits* of height n in $\mathcal{X}$ is a complete binary branching tree of height n, with each vertex labeled by a pair (A, O) with A a finite subset of X and O an orbit of the pointwise stabilizer in $\mathrm{Aut}(\mathcal{X})$S of A, such that the sets A increase as one moves along a path in the tree, and the two orbits lying below a given vertex are contained in the orbit attached to that vertex, and are pairwise disjoint.

2. The *orbit height* of $\mathcal{X}$ is the maximum value of n (or ∞) for which $\mathcal{X}$ has a tree of orbits of height n.

3. The structure $\mathcal{X}$ is *stable* if its orbit height is finite.

Smoothly approximable homogeneous structures with κ and r_κ finite are stable. The converse is an important structural result. Lachlan's theory consists of one part which provides a structural analysis and finiteness theorem for homogeneous structures in which not only κ and r_κ are fixed, but a bound on the orbit height is also fixed in advance. To complete the theory one must also bound the height in terms of κ and r_κ. We will look at this issue more closely in the next section.

5 The coordinatization theorem

Lachlan proposes to consider the following classification problems:

Given κ and r, classify the homogeneous structures $\mathcal{X}$ with $\kappa(\mathcal{X}) \leq \kappa$ and $r_\kappa(\mathcal{X}) \leq r$

$$(P_{\kappa,r})$$

It remains to be seen what constitutes a solution; in other words, what constitutes the specification of an infinite family of related structures. There are two approaches to this. For example, if the family in question is $\{m \cdot K_n\}$, disjoint sums of complete graphs of fixed size, we can pass to the limit structure, $\infty \cdot K_\infty$, and define the family in question as the set of finite approximations to the infinite limit – or we can look for the invariants m, n directly. It is reasonable to combine these approaches: identify the limit structures; show that there are finitely many; show that all sufficiently large structures approximate infinite limit structures; and identify the numerical invariants that control isomorphism types within each infinite family. Thus part of the work will take place "at infinity", in a model theoretic context, and part will take place in the "large finite", using combinatorics and permutation group theory. One speaks also of "shrinking" and "stretching"; shrinking an infinite structure to its finite approximations – which is relatively easy – and stretching a large finite structure to its infinite version (which amounts to giving an "explanation" of the finite structure as an approximation to an infinite one). To put the matter more briefly: there are finitely many infinite homogeneous structures of a given type, from which all sufficiently large finite homogeneous ones (and all their infinite limits) arise by a shrinking process. This is the basic finiteness result.

The notion of shrinking is given by smooth embedding: any finite structure which is smoothly embedded in a larger structure can be considered as a shrinking of the larger structure (it might be best at some later stage to exclude some very small smoothly embedded substructures). One can finesse the issue of stretching temporarily by deciding that a stretching of a structure is a structure in which it smoothly embeds, or in other words stretching is the reversal of shrinking; this is reasonable, but it just postpones the question of the existence of infinite stretchings, which is one of the main points: when can a finite structure be interpreted as a "template" for an infinite structure?

In any case, with this terminology, we may state:

Theorem 1 *Let $\sigma = (n_1, \ldots, n_r)$ be a type of relational system. There are finitely many homogeneous structures Γ_i of type σ such that every finite homogenous structure of type σ is obtained by shrinking one of the Γ_i.*

To put some flesh on these bones, it is necessary to look at numerical invariants. These are the analogs of "dimension" in vector spaces, but in the context of relational structures they are of a relatively degenerate type, as illustrated by the family of homogeneous graphs $m \cdot K_n$; here m is the number of classes of an invariant equivalence relation, and n is the size of the classes. This is close to the general case, for Lachlan's context.

A less trivial illustration is furnished by the graphs $\begin{bmatrix} n \\ k \end{bmatrix}$. Here there should be a single numerical parameter n (k is fixed, as it can be bounded in terms of κ). Evidently the parameter n is encoded by $\begin{bmatrix} n \\ k \end{bmatrix}$ with considerably more subtlety than in the case of $m \cdot K_n$. As mentioned in §3, we will think of $\begin{bmatrix} n \\ k \end{bmatrix}$ as a grassmannian structure associated with a degenerate geometry. As the examples in §3 may suggest, one would expect the geometries associated to a homogeneous structure to be degenerate (or of bounded size, hence more or less unproblematic). Accordingly we make the following definitions in general.

Grassmannians and invariants

1. A *coordinatizing structure* will be a structure Δ carrying an equivalence relation E with finitely many classes, such that the E-classes are permuted transitively by Aut Δ, and the group fixing the classes setwise is the product of the full symmetric groups on each class.

2. A *k-grassmannian* of Δ is the structure whose points are the subsets of Δ meeting each class in k points, whose full automorphism group is Aut Δ with the natural action. The invariant attached to a grassmannian is the size of each equivalence class in Δ.

3. The invariants attached to a homogeneous structure Γ are the invariants attached to all k-grassmannian structures which occur as primitive sections of Γ, with k at most half the size of each equivalence class in Γ.

The importance of these invariants can be seen in the following result from [CL].

Proposition 1 (Coordinatization) *For any specified type σ of relational structure, there is a bound m such that for every homogeneous relational structure Γ of type σ and any maximal (Aut Γ)-invariant equivalence relation E on Γ, one of the following holds*:

1. $|\Gamma/E| \leq m$; *or*

2. Γ/E *is a grassmannian of a coordinatizing structure.*

The proof of this relies heavily on permutation group theory, notably the O'Nan-Scott Lemma and the classification of the finite simple groups. Both the model theoretic content and the permutation group theoretic analysis are reviewed in [KL]. Stronger results in a similar vein are found in [KLM] and are essential to the development of theories of broader scope.

This result is the starting point for the study of the "variable" numerical invariants associated with a homogeneous structure. It should be clear enough how one expands or shrinks a grassmannian structure; to do either to a more general homogeneous structure requires more attention. Shrinking can be defined rather directly: one shrinks the coordinate structures and then sees which elements of the structure should be kept.

Stretching, on the other hand, is a problem. Indeed, the essence of the matter is to determine how large a dimension should be in order that it can be stretched freely, and this is a nontrivial question, very much at the heart of determining at what point the sporadic objects run out and the general case is encountered. The main result of [KL] is a clean approach to this problem, which we will indicate very briefly in the next section.

6 Amalgamation

If Γ is a homogeneous structure, then Sub(Γ) denotes the class of finite structures which are isomorphic with induced substructures of Γ. A countable homogeneous structure Γ is determined up to isomorphism by Sub(Γ). Furthermore the relevant classes of finite structures, that is, those which occur as Sub Γ with Γ homogeneous, are easily characterized by their intrinsic properties: closure under isomorphism and induced substructure, and the *amalgamation property*, which we now define.

Definition

1. An *amalgamation problem* is a triple of structures $\Gamma_0, \Gamma_1, \Gamma_2$ together with a pair of embeddings $\iota_i : \Gamma_0 \to \Gamma_i$ for $i = 1, 2$. A *solution* to

such a problem is a structure $\bar{\Gamma}$ and embeddings $\bar{\iota}_i : \Gamma_i \to \bar{\Gamma}$ so that $\bar{\iota}_1\iota_1 = \bar{\iota}_2\iota_2$.

2. A class $\mathcal{A}$ of structures has the *amalgamation property* if any amalgamation problem involving structures in the class has a solution in $\mathcal{A}$.

To see that Sub(Γ) has the amalgamation property for Γ homogeneous, note that we may take $\Gamma_0, \Gamma_1, \Gamma_2$ as substructures of Γ, and after applying suitable isomorphisms we may assume the embeddings $\iota_i : \Gamma_0 \to \Gamma_i$ are *inclusions*; in this case, set $\bar{\Gamma} = \Gamma_1 \cup \Gamma_2$ and let the $\bar{\iota}_i$ be inclusions as well.

Somewhat less evident, but straightforward nonetheless, is the fact that we can construct a homogeneous structure from any amalgamation class closed under isomorphism and induced substructure.

This gives us a quick avenue, in principle, to stretching. Given a large finite homogeneous structure Γ, let $\mathcal{A}$ be the class of all finite structures which embed "locally" into Γ (we will say in a moment what this means). If $\mathcal{A}$ is an amalgamation class, then the corresponding homogeneous structure should play the role of the "stretch" of Γ. To make this precise we define Sub(n, Γ) as the collection of finite structures such that every substructure of size at most n embeds into Γ. With n fixed, Sub(n, Γ) may be taken as the class of structures that embed locally into Γ. In [KL] the following is proved, as Theorem 9.2:

Theorem *For each fixed type σ of relational structure, there is a fixed N such that for every homogeneous finite Γ of type σ, there is some $n \leq N$ for which* Sub(n, Γ) *is an amalgamation class, and for which the corresponding homogeneous structure is smoothly approximable.*

Obviously this sketch leaves out not only all of the proofs, but a number of highly relevant statements and definitions, particularly relating to the treatment of the numerical invariants in the preceding section and their relation to the theory of stretching. We refer the reader to [KL] for more details and further references, as well as a discussion of effectivity.

7 Rank: stability and bounds

We defined the *orbit height* of a permutation group in §4 and we called a homogeneous structure *stable* if its orbit height is finite.

For example, the orbit height of a set on which the full symmetric group acts is 1, since one cannot split an orbit into two orbits of size greater than 1. On the other hand the orbit height of the rational order is infinite, since one can split an infinite interval (a, b) into two such, and repeat. The orbit height is a notion of "dimension" (a measurement of noetherianity). The reader can test this stable/unstable dichotomy on the explicit list of homogeneous graphs given in §4.

The usual definitions of stability in model theory are of a more general character, but reduce to this notion in the homogeneous case. To some extent our initial presentation put the cart before the horse. In the most satisfactory version of Lachlan's theory, the starting point is stability. One proves eventually:

Proposition 2 *Within the class of homogeneous structures of a fixed type, the stable ones are those which can be smoothly approximated by finite homogeneous structures of the same type.*

Lachlan's theory gives the classification of all stable homogeneous structures of a fixed type, which includes the classification of the finite ones, but which makes use of the structure of the infinite ones along the way: this class falls into finitely many families, each parametrized by finitely many numerical invariants; and when all invariants are made finite, the resulting structure is finite.

The result on coordinatization (Proposition 1) is actually *equivalent* to the existence of a uniform bound on the orbit heights for stable homogeneous structures of fixed type, though one needs a good dose of model theory to see this. In any case, in all versions of this theory to date one gets the coordinatization result from group theory (at the finite level) and this then allows the introduction of model theoretic techniques to handle the infinite limits, returning, eventually, to the large finite structures. Thus our understanding of the infinite structures requires information coming from their finite approximations, but at the same time our understanding of the finite structures depends on a consideration of their infinite limits.

For unstable homogeneous structures, as we have noted earlier, we have explicit classifications in some nontrivial cases, but no theory.

8   Smoothly approximable structures

Lachlan's theory can be extended *mutatis mutandis* to smoothly approximable structures, and if one does so then the supply of geometries is enlarged to include all of the classical geometries (in both affine and projective flavors) which are so visibly absent in the homogeneous case. This is one of the most salient differences. It would be easy enough to carry this aspect along through the theory, but eventually the less trivial geometric structure in these geometries has a certain impact on developments and a more sophisticated dose of model theory comes into play, modeled on standard developments in stability theory. As it happens, the geometries involved are not in fact stable in general, which is one indication that some price is to be paid for the generalization. This is discussed in [Hr, Ch].

Smooth approximability is a somewhat peculiar hypothesis (as a point of departure) from the point of view of model theory, but Lachlan observed that in view of the central role it plays in the theory of stable homogeneous structures and related developments, it is reasonable to work in this category, discarding any more special model theoretic hypotheses. Furthermore dealing with this class amounts to dealing at the finite level with permutation groups with a bound on the number of orbits on 4-tuples; and though one might have preferred to replace 4 by 2 here, this is certainly a natural class to work with.

It is not at all clear *a priori* that the only geometries which are relevant here are classical ones; and strictly speaking, this is not even true, as in characteristic 2 there is another family of geometries falling just outside the classical camp. One can read off the relevant list of geometries from the explicit classification of the large finite primitive structures with a bounded number of orbits on 4-tuples [KLM as modified in Mp]. It is fortunate that permutation group theory is so effective on primitive structures while model theory has good tools for reducing imprimitive structures to primitive ones.

We will mention two of these geometries, to give an indication of the sort of objects that come into play at this level.

Example 9: polar geometry   One has a pair (V, V^*) with V an infinite dimensional vector space over a finite field (of countably infinite dimension) and V^* a dense subspace of its dual. This is the smooth limit of analogous finite dimensional structures in which V^* is the full dual of V.

In model theory we consider the pair (V, V^*) as a single geometry. One of the complications that arises is that in encountering V embedded in a larger structure one cannot easily tell whether it occurs as a subspace with no additional structure, or as half of a polar pair. For example, we might have V together with some grassmannian associated with V^*; we would then have to reconstruct V^* from its grassmannian in order to have the second component of the polar pair. None of this is terribly troublesome; it just needs to be dealt with. One of the basic ideas of pure model theory, expressed by the theory of "orthogonality", is that *geometries interact trivially*: either they can be identified (with some deformation), or they are unrelated. If V and V^* are taken to be *separate* geometries, this orthogonality principle is lost; but if such things are not allowed, the principle of orthogonality can be saved.

Example 10: The quadratic geometries There is a family of geometries in characteristic 2 which appears to blend some of the features of affine and projective geometry. It arises because orthogonal groups are contained in symplectic groups in characteristic 2.

The underlying set Q of this geometry in a given (even) dimension is defined as follows: let V be a symplectic space of dimension $2n$, with inner product $(\ ,\ )$, and let Q be the set of all quadratic forms q for which we have $q(x + y) = q(x) + q(y) + (x, y)$.

There is a regular action of V^* on Q (or of V on Q after identifying V and V^*): $q.\lambda = q + \lambda^2$ for $q \in Q$, $\lambda \in V^*$. In this sense (but only in this sense) the space looks affine. The stabilizer of a point q is of course the associated orthogonal group $O(q)$. The same structure exists in an infinite dimensional version but there is one delicate point: the witt defect of $q \in Q$ is well defined when the dimension is finite, and in passing to a *smooth limit* a formal witt defect function ω is inherited, though the witt defect itself is no longer very meaningful. (The equivalence relation defined by "same witt defect" can be defined intrinsically from the geometric structure; only the value of the witt defect is lost.) The upshot of all this is that in the infinite limit, one takes Q together with V^* and its regular action, as well as the structure on V^*, together with the formal witt defect.

This is the least classical of the geometries that come into play here.

9 $\kappa(\mathcal{X})$: problems

The question naturally arises: for one's favorite *primitive* structures, how is $\kappa(\mathcal{X})$ computed in practice? Much of what is understood about this was reviewed in §3. We comment here on cases we do not understand. There are tools for the general analysis of primitive permutation groups, beginning with the O'Nan-Scott lemma and continuing in work of Aschbacher, and among other things, one would like to know how κ behaves relative to this. Some general theory for wreath products is found in [CMS] with reasonably satisfying results, some loose and very general, others more detailed under rather specific hypotheses. The following test problem, which remains wide open, is my personal favorite in this area:

Problem 1 Determine the finite primitive binary structures.
$(\kappa(\mathcal{X}) = 2)$

Model theorists might be tempted by an indirect inductive approach involving imprimitive binary structures as well, but this does not seem very promising, and the group theoretic approach may work here. Analogously one might ask for the classification of all sufficiently large primitive structures with a specified bound on $\kappa(\mathcal{X})$. The invariant κ may behave well enough – in terms of crude lower bounds – to make an O'Nan-Scott type of analysis feasible. Certainly κ is not well behaved under restriction to the socle, as one sees by comparing $\mathrm{Sym}(n)$ and $\mathrm{Alt}(n)$, but since only lower bounds are needed, the situation may possibly be manageable.

In §3 we mentioned the conjectured answer in the binary case: oriented and unoriented cycles of prime order; a naked set (with equality); and the peculiar edge-colored graphs associated with primitive homogeneous 1-dimensional affine groups which are not strictly linear, namely those of the form $\Gamma_q = \mathbb{F}_{q^2} \rtimes \mu_{q+1} \rtimes \mathbb{Z}/2\mathbb{Z}$, acting on the base field.

Problem 2   Let $\mathcal{E}_{k,l}$ be the set of partitions of a set of $n = k \cdot l$ elements into k classes of size l. Estimate $\kappa(\mathrm{Sym}(n), \mathcal{E}_{k,l})$, and consider more general partition types, as well as the action of $\mathrm{Alt}(n)$.

One can get some partial results using estimates for κ of a wreath product of actions on k-sets, though unfortunately in this connection one also needs the values of κ for wreath products falling in the difficult range, such as those handled by Saracino for $k = 1$.

For example we have:

$$\text{With } l_0 < l/2, \ \kappa(\mathcal{E}_{k,l}) \geq \kappa(\mathcal{E}_{2,l}) \geq \kappa\left(\begin{bmatrix} l \\ l_0 \end{bmatrix}^2\right)$$

For the second inequality, work with the stabilizer of a single point in $\mathcal{E}_{2,l}$. An example: $\kappa(\mathcal{E}_{2,4}) \geq \kappa(4^2) = 4$; the exact value in this case is 5.

One can also show:

$$\kappa(\mathcal{E}_{n,2}) \geq n$$

using a form of the "Möbius band" example of Lachlan which was given at the end of [CMS] in a similar context. This lower bound may possibly be the correct value.

Problem 3   Improve the estimates on $\kappa\left(\begin{bmatrix} n \\ k \end{bmatrix}^d\right)$ in the range $n \leq 2k([\log_2 d] + 1)$, getting exact values or at least asymptotically accurate estimates.

The results of [CMS] and the extraordinarily precise analysis by Saracino in the case $k = 1$ suggest that the best way to approach this is in terms of a function $\delta_k(r, n)$, with the following subtle definition, which will be elucidated momentarily: $\delta_k(r,n)$ is the least d (or ∞ if none exists) for which there are two multisets $\mathcal{H}$, $\mathcal{H}'$ of r-labeled k-uniform hypergraphs on n vertices whose $(r-1)$-restrictions coincide up to isomorphism. Here a multiset is a set with multiplicities; an r-labeled k-uniform hypergraph on n vertices is a map λ from $\{1,\ldots,r\}$ to the k-subsets of an n-element set; an A-restriction of an r-labeled hypergraph corresponding to a subset A of $\{1,\ldots,r\}$ is the A-labeled hypergraph obtained by restricting λ to A; and an $(r-1)$-restriction is an A-restriction with $|A| = r-1$. Finally, we say that the A-restrictions of $\mathcal{H}$ and $\mathcal{H}'$ coincide if there is a bijection between $\mathcal{H}$ and $\mathcal{H}'$ so that corresponding hypergraphs have isomorphic A-restrictions, where isomorphisms are permitted to permute the n vertices but the domain A is fixed; and we say that the $(r-1)$-restrictions coincide, if for each A of cardinality $r-1$, the A-restrictions coincide. Examples are in order.

For $k = 1$, an r-labeled hypergraph is a function from $\{1,\ldots,r\}$ to $\{1,\ldots,n\}$, and the isomorphism types (allowing the action of $\mathrm{Sym}(n)$) are classified by equivalence relations on $\{1,\ldots,r\}$. Thus in this case $\mathcal{H}$ and $\mathcal{H}'$ can be viewed more simply as collections of equivalence relations on $\{1,\ldots,r\}$ with at most n classes. For example for $n = 2$ and r even, we may let $\mathcal{H}$ consist of the partitions of $\{1,\ldots,r\}$ into two pieces of even size (one may be empty); and let $\mathcal{H}'$ consist of the partitions into two pieces of odd size. One may then check directly that the A-restrictions, for $|A| = r-1$, consist of all equivalence relations on A, each occurring once. This shows $\delta_1(r,2) \leq 2^{r-2}$ and further analysis shows this estimate is exact. For more on δ_1 see §3.

Little is known about δ_2, but some examples may clarify the meaning of the definition.

$\delta_k(r,n) = \delta_{n-k}(r,n)$ for any k,n since there is a canonical idenitification between k-sets and their complements. In particular $\delta_2(r,3) = \delta_1(r,3)$. Therefore we will consider the case $n = 4$.

Example $\delta_2(4,4) = 2$. We give an explicit example. Let a,b,c,d be the four edges of a 4-cycle and define $\mathcal{H}$, $\mathcal{H}'$ as follows (for each r-labeled graph we just list the values of $\lambda(1),\ldots,\lambda(4)$ in order:

$\mathcal{H}$: (1) $a/b/c/d$; (2) $a/b/a/b$;
$\mathcal{H}'$: (1) $a/b/c/b$; (2) $b/a/b/c$.

This explicit example shows $\delta_2(4,4) \leq 2$; evidently $\delta_2(4,4) > 1$.

Example $\delta_2(5,4) \leq 15$.

Let Λ be the set of injective functions from $\{1,2,3,4,5\}$ into K_4; these may also be thought of as labelings of K_4 minus an edge by distinct labels $1, \ldots, 5$. Sym(6) and Sym(4) act naturally on the edges and vertices; Sym(4) preserves isomorphism types and Sym(6) acts regularly on Λ. Thus the isomorphism types represented by Λ may be identified with the coset space Sym(6)/Sym(4) under a natural embedding of Sym(4) into Sym(6) which actually takes Sym(4) into Alt(6). In particular this space falls naturally into even and odd types (though the determination of which is which is of course arbitrary). Let $\mathcal{H}$ and $\mathcal{H}'$ be representatives of these two classes. Then $\mathcal{H}$ and $\mathcal{H}'$ each consist of 15 5-labeled graphs on 4 vertices, and we claim that the restrictions of $\mathcal{H}$ and $\mathcal{H}'$ obtained by deleting any one label coincide. For each label i there is a natural bijection between $\mathcal{H}$ and $\mathcal{H}'$ in which each 5-labeled graph is replaced by the corresponding 5-labeled graph in which the label i is moved to the unlabeled edge. In each case the same thing can be accomplished by a transposition in Sym(6), so this switches the even and odd types, and it obviously preserves the i-restrictions.

This example shows that $\delta_2(5,4) \leq 15$. Full information on $\delta_2(r,n)$ for a fixed value of n determines the value of $\kappa(\left[\begin{smallmatrix} n \\ 2 \end{smallmatrix}\right]^d)$ for all d. For example, our estimate for $\delta_2(5,4)$ suggests that $\kappa(\left[\begin{smallmatrix} 4 \\ 2 \end{smallmatrix}\right]^d)$ may be 4 for $2 \leq d \leq 14$ and 5 for $d = 15$, but to pin this down one would need not only the exact value of $\delta_2(5,4)$, but a little more information about $\delta_2(r,4)$ in general.

References

[As]   M. Aschbacher, *The Theory of Finite Groups*, Cambridge Studies in Advanced Mathematics 10, Cambridge University Press, 1986. ix+274 pp.

[Ca1]   P. Cameron, Finite permutation groups and finite simple groups, *Bull. London Math. Society* **13** (1981), 1–22.

[Ca2]   P. Cameron, *Oligomorphic Permutation Groups*, London Mathematical Society Lecture notes 152, Cambridge University Press, 1990.

[Ch]   G. Cherlin, Large finite structures with few types, in *Proceedings of NATO ASI*, Fields Institute, Toronto, August 1996, NATO ASI Series C **496**, Kluwer, Dordrecht, 1997, 53–105.

[CL]   G. Cherlin and A. Lachlan, Stable finitely homogeneous structures, *TAMS* **296** (1986), 815–850.

[CMS] G. Cherlin, G. Martin, and D. Saracino, Arities of permutation groups: Wreath products and k-sets, *J. Combinatorial Theory*, Ser. A **74** (1996), 249–286.

[DM] J. Dixon and B. Mortimer, *Permutation Groups*, Graduate Texts in Mathematics, Springer, New York, 1996.

[Ga] A. Gardiner, Homogeneous graphs, *J. Comb. Theory Series B* **20** (1976), 94–102.

[GK] Ya. Gelfand and M. Klin, On k-homogeneous graphs, in *Algorithmic Studies in Combinatorics*, 76–85, Nauka, Moscow, 1978.

[Hr] E. Hrushovski, Finite structures with few types, in *Finite and Infinite Combinatorics in Sets and Logic*, eds. N. W. Sauer, R. E. Woodrow, and B. Sands, NATO ASI Series C vol. 411, Kluwer, Dordrecht, 1993.

[KLM] W. Kantor, M. Liebeck, and H. Macpherson, $\aleph_0$-categorical structures smoothly approximated by finite substructures, *Proceedings of the London Mathematical Society*, Series 3 **59** (1989), 439–463.

[KL] J. Knight and A. Lachlan, Shrinking, stretching, and codes for homogeneous structures, in *Classification Theory*, J. Baldwin ed., LNM 1292, Springer, New York, 1985.

[La1] A. Lachlan, On countable stable structures which are homogeneous for a finite relational language, *Israel J. of Math.* **49** (1984), 69–153.

[La2] A. Lachlan, Homogeneous structures, in *Proceedings of the International Congress of Mathematicians, Berkeley 1986*, A. Gleason ed., v. 1, 314–321, American Mathematical Society, Providence, R.I., 1987.

[LW] A. Lachlan and R. Woodrow, Countable ultrahomogeneous undirected graphs, *Transactions of the American Mathematical Society* **262** (1980), 51–94.

[Mp] H. Macpherson, Homogeneous and smoothly approximated structures, in *Proceedings of NATO ASI*, Fields Institute, Toronto, August 1996, NATO ASI Series C **496**, Kluwer, Dordrecht, 1997.

[Sa] D. Saracino, The arity of n^d, two preprints, (1996) and manuscript in preparation (1997).

[Sh]   J. Sheehan, Smoothly embeddable subgraphs, *J. London Math. Soc.* **9** (1974), 212–218.

Department of Mathematics
Hill, Center, Busch Campus
Rutgers University
New Brunswick, NJ 08903
E-mail: cherlin@math.rutgers.edu

AMS Subject Classification: Primary 03C60, 20B25; Secondary 03C13

Three Questions about Simplices in Spherical and Hyperbolic 3-Space

*Johan L. Dupont and Chih-Han Sah**

Abstract

We investigate the relations among three questions raised respectively by Hilbert, by Cheeger–Simons and by Milnor.

Introduction

We begin with a rapid review of the Third Problem of Hilbert [Hilbert 1900], which originated with Gauss [Gauss, 1844]. Two Euclidean polyhedra P and Q are said to be scissors congruent if P and Q can each be decomposed into a *finite* pairwise interior disjoint union of polyhedra P_i, Q_i, $1 \leq i \leq n$, so that P_i is congruent to Q_i for each i. They are said to be stably scissors congruent if we can find congruent polyhedra R and S, such that the interior disjoint union of P and R is scissors congruent to the interior disjoint union of Q and S. As it turned out, under fairly general conditions, Zylev [Zylev, 1965, 1968] showed that there is no difference between these two concepts. As the third of the list of twenty-three problems, Hilbert asked: Show that there exist two tetrahedra with the same base and the same height that are not scissors congruent. This was promptly solved by his student M. Dehn [Dehn, 1901]. Thus, Dehn showed that the tetrahedron with vertices at $(0,0,0)$, $(1,0,0)$, $(0,1,0)$, $(0,0,1)$ and the tetrahedron with vertices at $(0,0,0)$, $(1,0,0)$, $(1,1,0)$, $(1,1,1)$ are not scissors congruent. In fact, there is a continuum number of such examples.

This undoubtedly placed a damper on further investigations for quite a while. To accomplish this, Dehn extended the definition of invariants, now called Dehn invariants, that were first considered in [Bricard, 1896] in a more restricted setting. It is then easy to show that the cube has

*For both authors, work was supported in part by grants from Statens Naturvidenskabelige Forskningsråd and the Gabriella and Paul Rosenbaum Foundation.

zero Dehn invariants, while the regular tetrahedron has non-zero Dehn invariants. In the work of Dehn, it was quite difficult to show that these invariants in fact depended only on the scissors congruence classes of a polyhedron.

An elegant exposition was given by H. Hopf in 1946 at NYU, reprinted in [Hopf, 1983]. The modern formulation of the Dehn invariant is first found in [Jessen, 1968] in terms of tensor product of the additive abelian groups $\mathbb{R}$ and $\mathbb{R}/\mathbb{Z}\pi$. If Δ is convex polyhedron, then its Dehn invariant is the sum over all the edges the elements $\ell \otimes \theta$ where ℓ denotes the length of the edge and θ denotes the interior dihedral angle along this edge. In his fundamental work, Jessen simplified and made precise the tour-de-force solution in [Sydler, 1965] for the converse of the Theorem of Dehn leading to the introduction of homological algebraic techniques, see [Jessen–Karpf–Thorup, 1968]. Alternate solutions extending this homological theme can now be found in [Dupont–Sah, 1990], [Cathelineau, 1998]. Thus, we restate the original theorem:

Theorem 1 (Dehn–Sydler) *Two Euclidean polyhedra are scissors congruent if and only if they have the same volume and the same Dehn invariant.*

Theorem 1 was extended to dimension 4 in [Jessen, 1972] but remains open in all higher dimensional Euclidean spaces. In [Dehn, 1901, 1906], Dehn invariants were extended to hyperbolic and spherical 3-spaces. By a non-constructive counting argument, Dehn showed the existence of non-Euclidean tetrahedra with same volume but different Dehn invariants. In view of these results, it is reasonable to ask:

Generalized Third Problem of Hilbert. *Is it true that two geodesic polyhedra in spherical (respectively hyperbolic) 3-space are scissors congruent if and only if they have the same volume and the same Dehn invariant?*

The same question can be raised in higher dimension spaces as well. The definition of Dehn invariants extends in a variety of ways to all dimensions; in particular, at the level of scissors congruence groups, the generalized Dehn invariants become the structure maps for a comodule structure with respect to suitable Hopf algebras over the integers; see [Sah, 1979; Goncharov, 1999]. In a conference held at Stanford in 1973, J. Cheeger and J. Simons raised the following question, see [Cheeger–Simons, 1985]:

Rational Simplex Problem (Cheeger and Simons). *Given a geodesic simplex in spherical 3-space so that all of its interior dihedral* *angles are rational multiples of* π, *is it true that its volume is a rational multiple of the volume of the 3-sphere (namely,* $2\pi^2$)?

Their conjecture [Cheeger–Simons, 1985] is that the answer should be negative "most" of the time. However, in all of the known cases, for example, see [Coxeter, 1935], the answer is yes. This is because the cases considered by Coxeter can all be shown to be scissors congruent to fundamental domains of finite groups acting on the 3-sphere, so that their volumes must necessarily be rational multiples of the volume of the 3-sphere. In fact, all the exact volumes known to us arise from the elementary geometric techniques of scissors congruences, in other words, through finite cutting and pasting along geodesic faces.

As a part of the program to study the geometry and topology of 3-dimensional manifolds, see [Thurston, 1978–79], Milnor studied the Lobachevsky function Л [Milnor 1980]. Consider the dilogarithm function $L_2(z) = \sum_{n \geq 1} z^n/n^2$ studied by Euler, Abel and many others, see [Lewin, 1958, 1981, 1991]. It is related to the Lobachevsky function

$$\text{Л}(\theta) = -\int_0^\theta \log |2 \sin t| \, dt, \tag{1}$$

by the formula

$$L_2(e^{2\iota\theta}) = \pi^2/6 - \theta(\pi - \theta) + 2\iota\text{Л}(\theta), \quad 0 \leq \theta \leq \pi. \tag{2}$$

If we use the Beltrami upper half space model of the hyperbolic 3-space so that its ideal boundary is identified as the projective line over the complex numbers, then $\text{Л}(\theta)$ is one half of the volume of a geodesic hyperbolic 3-simplex $(\infty, 0, 1, e^{2\iota\theta})$ with infinite vertices at ∞, 0, 1, and $e^{2\iota\theta}$, where $0 \leq \theta \leq \pi/2$.

$\text{Л}(\theta)$ is known to be an odd, real analytic periodic function with period π and achieve a unique maximum on $[0, \pi/2]$ at $\pi/6$. Its graph resembles that of $\sin 2\theta$ and it satisfies the following functional identity for every integer $n \geq 0$; see [Milnor 1980]:

$$\text{Л}(n\theta) = n \sum_{k \bmod n} \text{Л}(\theta + k\pi/n), \quad n > 0 \text{ is an integer.} \tag{3}$$

Milnor's conjecture. *Consider only angles* θ *that are rational multiples of* π; *then every rational linear relation of the form*

$$\sum q_j \text{Л}(\theta_j) = 0, \quad q_j \text{ rational,} \tag{4}$$

is a consequence of the relations

$$Л(\theta + \pi) = Л(\theta), \quad Л(-\theta) = Л(\theta), \quad \text{and} \tag{5}$$

$$Л(n\theta) = n \sum_{k \bmod n} Л(\theta + k\pi/n). \tag{6}$$

In other words, for each integer $n > 2$, the values $Л(k\pi/n)$, should span a $\mathbb{Q}$-vector subspace of dimension $\phi(n)/2$ in the $\mathbb{Q}$-vector space $\mathbb{R}$ of real numbers, where ϕ denotes the Euler totient function. To be precise, a $\mathbb{Q}$-basis is formed by restricting k to satisfy

$$0 < k < n/2, \quad \gcd(k, n) = 1. \tag{7}$$

Without loss of generality, we may assume n is even. We note that $\phi(n)/2$ is the degree of the totally real subfield of the n-th cyclotomic field $\mathbb{Q}[\exp(2\pi\iota/n)]$ fixed by complex conjugation. By Dirichlet's Unit Theorem, the group of units in the ring of integers $\mathbb{Z}[\exp(2\pi\iota/n)]$ is the direct sum of the cyclic group of n-th roots of 1 and a free abelian group of rank $\phi(n)/2 - 1$. When we pass to the totally real subfield, the rank of the free part does not change.

The purpose of the present paper is to point out the relations among these three questions. In particular, we show that all three of these questions are quite deep. It should be noted that there are several interesting works of Walter Neumann and Jun Yang [Neumann–Yang, 1995, 1999] which also contains a number of references in a variety of directions.

1   Main results

In both spherical and hyperbolic 3-spaces, we provide explicit examples to illustrate the relations. We begin with spherical 3-space. A *lune* is defined to be the orthogonal suspension to the two poles of a geodesic spherical triangle lying on the equator of spherical space. This definition will be made more precise later. It is sufficient to note that the volume of such a lune is $\pi \cdot (\alpha + \beta + \gamma - \pi)/2$, where α, β, and γ denote the vertex angles of the spherical triangle. An *orthoscheme* is the generalization of a right triangle. Thus, the vertices may be ordered as A, B, C, and D so that the edge BC is perpendicular to the edge AB and the edge CD is perpendicular to the geodesic plane of the triangle ABC.

Theorem 2 *Let Δ denote a spherical simplex with all of its dihedral angles in $\mathbb{Q}\pi$. Then, we have the following alternatives:*

(a) Δ is scissors congruent to a lune so that its volume is in $\mathbb{Q}\pi^2$.

(b) Δ is not scissors congruent to a lune and exactly one of the following possibilities holds:

   (b₁) Δ leads to a negative answer of the generalized Hilbert Third Problem for spherical 3-space; or

   (b₂) Δ does not have volume in $\mathbb{Q}\pi^2$, so it is an example desired by Cheeger–Simons.

Furthermore, for an orthoscheme Δ, case (a) and (b) are distinguished by an explicit algorithm.

In either case (b), we have infinitely many explicit examples for the original Hilbert Third Problem in spherical 3-space.

In the case of hyperbolic 3-space, we do not have any obvious analogue of the 3-sphere. Nevertheless, we have a similar statement. For this, we need to recall the definition of a totally asymptotic 3-simplex in hyperbolic space. In the upper half space model, we take any three noncollinear points of the Gaussian z-plane. The vertical rays starting from points of the Gaussian plane are the geodesics joining these points to the point at infinity ∞. The remaining geodesics are half circles in planes perpendicular to the Gaussian z-plane. If one of the vertices of a totally asymptotic 3-simplex is at ∞, then, using vertical projection to the Gaussian z-plane, we obtain a Euclidean triangle. The three vertices of this triangle are then the other three vertices of our totally asymptotic hyperbolic 3-simplex. The congruence class of this 3-simplex is the same as the similarity class of this Euclidean triangle. In particular, the three dihedral angles along the three edges meeting at ∞ are the corresponding vertex angles of the Euclidean triangle. This is the model used in [Milnor, 1980] to obtain the volume formula. As shown in [Lemma 4.4, Dupont–Sah, 1982], the Dehn invariant of this totally hyperbolic 3-simplex $\Delta(\alpha, \beta, \gamma)$ is

$$\log(2\sin\alpha)\otimes\alpha + \log(2\sin\beta)\otimes\beta + \log(2\sin\gamma)\otimes\gamma \in \mathbb{R}\otimes\mathbb{R}/\mathbb{Z}\pi. \quad (8)$$

We note the similarity between (8) and (1).

Theorem 3 *Let $(\infty, 0, 1, e^{2\pi\iota/n})$, $n > 6$ denote the totally asymptotic hyperbolic 3-simplex with vertices at $\infty, 0, 1$ and $e^{2\pi\iota/n}$. Let $1/6 < \theta < 1/2$ denote the unique real number so that:*

$$Л(\theta\pi) = Л(\pi/n)$$

Then, one of the following alternatives holds:

(a) θ is irrational, hence $(\infty, 0, 1, e^{2\pi\iota\theta})$ has non-zero Dehn invariant and we have an explicit pair of hyperbolic 3-simplices with the same volume but different Dehn invariant as desired by Dehn;

(b) θ is rational so that Milnor's conjecture is false.

We again have an infinite number of explicit examples for the original Hilbert Third Problem in hyperbolic 3-space. However, in this case, we do not know how to determine if θ is irrational or not.

2   Gram parametrization of n-simplices in classical geometries

For the convenience of the reader, we provide a review of the parametrization of geodesic simplices by their Gram matrices, see also [Milnor, 1980].

In each of the three classical geometries of dimension n — hyperbolic, Euclidean, spherical, a geodesic n-simplex is defined up to isometry by its unordered set of $n + 1$ vertices. In the case of hyperbolic geometry of dimension greater than one, we have the added freedom in extending the vertices to infinity and still retain the finiteness of the corresponding volume. Such simplices will be called k-asymptotic if exactly k of its vertices are infinite. When $k = n + 1$, we speak of totally asymptotic n-simplices. These are also called ideal simplices. For $n > 1$, it is known [Haagerup–Munkholm, 1981] that the regular totally asymptotic n-simplex achieves the unique maximum volume among all the geodesic n-simplices in hyperbolic n space. Its codimensional 2 dihedral angle θ_n is acute and cos $\theta_n = 1/(n-1)$. Thus, θ_n is a rational multiple of π exactly when $n \leq 3$; in particular, $\theta_2 = 0$ was known to Gauss, see [Milnor, 1982] for a wealth of information.

We now review the parametrization of n-simplices Δ by their Gram matrices $G(\Delta)$. In each case, we may embed the geometry of dimension n isometrically into the next higher dimension $n + 1$. Each isometry can then be extended to an isometry in two different ways to the next higher dimension. More generally, the extension involves an arbitrary element of the orthogonal group in the normal direction. In the geometry itself, the isometry group acts transitively on the flags (increasing sequences) of subspaces. Thus, for any n-simplex Δ with vertices v_i, we can define exterior unit vectors u_i to the hyperplane face spanned by the vertices other than v_i. Because of the homogeneity of the isometry group, it does not matter where we choose to locate this unit vector. In particular, it is possible to define the exterior dihedral angle between two hyperplane faces to be the angle between u_i and u_j. The supplementary angle $\theta_{i,j}$ is then the interior dihedral angle between these two codimensional one faces. As we shall see later, for $n > 1$, a spherical and a hyperbolic n-simplex is determined up to isometry by this collection of dihedral angles, while an Euclidean n-simplex is determined up to similarity by these dihedral angles.

We begin with an $\mathbb{R}$-vector space $\mathbb{V}(\epsilon)$ of dimension $n + 1$ and $\mathbb{R}$-basis e_i, $0 \leq i \leq n$, where $\epsilon = -1$, 0, 1 corresponds respectively to hyperbolic, Euclidean, and spherical geometry. We define the inner product $\langle \cdot, \cdot \rangle_\epsilon$ on $\mathbb{V}(\epsilon)$ so that

$$\begin{aligned} < e_i, e_j >_\epsilon &= 0, & i \neq j; \\ < e_0, e_0 >_\epsilon &= \epsilon, & \text{and} \\ < e_i, e_i >_\epsilon &= 1, & i > 0. \end{aligned} \tag{9}$$

Each of the three geometries can be modelled by a set of rays of the form $\mathbb{R}^+ v$ in $\mathbb{V}(\epsilon)$.

In the case of spherical n-space, there is no restriction. Each ray contains a unique vector v with $\langle v, v \rangle_1 = 1$. The vertices of a spherical n-simplex Δ is then determined by a collection of $n+1$ $\mathbb{R}$-linearly independent unit vectors, say v_i. We can then define, in a unique way, a "dual" simplex, Δ^* with vertices u_i according to the following rules:

$$\langle u_i, v_j \rangle_1 = 0, \ i \neq j, \ \langle u_i, v_i \rangle_1 < 0, \ \langle u_i, u_i \rangle_1 = 1. \tag{10}$$

The spherical n-simplex Δ is defined by the following inequalities:

$$\Delta = \{ v \in \mathbb{V}(1) \mid \langle v, v \rangle_1 = 1, \ \langle v, u_i \rangle_1 \leq 0 \}. \tag{11}$$

The $(n + 1) \times (n + 1)$ matrix $G(\Delta) = (\langle u_i, u_j \rangle_1)$ is called the *Gram matrix* associated to the spherical n-simplex Δ and has the following properties:

$$\Delta = (\Delta^*)^* \quad \text{and} \quad G(\Delta^*) = (\langle v_i, v_j \rangle_1). \tag{12}$$

$$G(\Delta) \text{ is an} (n + 1) \times (n + 1) \text{ real symmetric matrix.} \tag{13}$$

$$G(\Delta) \text{ has diagonal entries equal to 1.} \tag{14}$$

For $0 \leq i \neq j \leq n$, the (i, j)-th entry of $G(\Delta)$ is equal to $-\cos(\theta_{i,j})$, and $\theta_{i,j} =$ the interior dihedral angle between the codimensional 1 faces opposite v_i and v_j. $\tag{15}$

Permutation of the vertices of Δ corresponds to simultaneous column and row permutations of $G(\Delta)$. $\tag{16}$

The $n \times n$ principal minors of $G(\Delta)$ are positive definite. $\tag{17}$

$$\det(G(\Delta)) > 0. \tag{18, +}$$

For the sake of convenience, we also state the conditions corresponding to the Euclidean and the hyperbolic cases:

$$\det(G(\Delta)) = 0. \tag{18, 0}$$

$$\det(G(\Delta)) < 0. \tag{18, -}$$

According to the condition (18), we will speak of a Gram matrix of spherical, Euclidean or hyperbolic type. In each case, Δ is compact.

It is not difficult to check that any matrix with properties (13)–(17) and (18, +) defines a spherical n-simplex up to isometry. To be precise, we use the matrix to define an inner product on an abstract $\mathbb{R}$-vector space of dimension $n + 1$ with basis v_i and invoke Sylvester's Law of Inertia to conclude that this inner product space is isomorphic to $\langle \cdot, \cdot \rangle_1$ on $\mathbb{V}$ so that we can define Δ up to isometry by (11). If $\epsilon_i = \pm 1$ are chosen at will, then we get a total of 2^{n+1} spherical n-simplices with vertices $\{\epsilon_i v_i \mid 0 \leq i \leq n\}$. The corresponding Gram matrix is obtained from $G(\Delta)$ through simultaneous multiplication of the i-th column and i-th row by ϵ_i. These 2^{n+1} spherical n-simplices form a pairwise interior disjoint partition of the unit n-sphere. If we use sccl to denote the spherical convex closure of a set of $n + 1$ $\mathbb{R}$-linearly independent unit vectors, then the union of the two simplices:

$$\Delta = \mathrm{sccl}\{v_0, \cdots, v_n\} \quad \text{and} \quad \Delta_i' = \mathrm{sccl}\{v_0, \ldots, -v_i, \ldots, v_n\} \tag{19}$$

forms a *lune*. It is the orthogonal suspension of the $(n - 1)$-simplex in the orthogonal complement of $\mathbb{R}v_i$ whose dual is the $(n - 1)$-simplex with vertices u_j, $j \neq i$.

Evidently, positive definite Gram matrices are inductive in the sense that all of its principal minors are again positive definite. In terms of $G(\Delta)$, the unit vector u_i is the exterior unit normal to the codimensional one face of Δ spanned by the vertices v_j with $j \neq i$. In particular, the unit vector u_i can be placed at the vertex v_i to provide the picture of the exhaust plume at the tail v_i of a rocket.

We next identify the Euclidean n-space $\mathbb{R}^n$ with the *subset* of $\mathbb{V}(0)$:

$$\mathbb{R}^{0,n} = e_0 + \mathbb{R}^n = e_0 + \sum_{i>0} \mathbb{R}e_i \tag{20}$$

together with the induced inner product $\langle \cdot, \cdot \rangle_0$. Thus, we send v in $\mathbb{R}^n$ to $e_0 + v$ in $\mathbb{R}^{0,n}$. Each such v then corresponds to the unique open ray $\mathbb{R}^+(e_0 + v)$ in $\mathbb{V}(0)$. The main advantage of this model is that the group of

Euclidean motions of the n-dimensional Euclidean space can be identified with the subgroup of all the $\langle \cdot, \cdot \rangle_0$-orthogonal linear transformations of our $(n+1)$-space that stabilizes $\mathbb{R}^{0,n}$. Moreover, under this identification, $n+1$ points of $\mathbb{R}^n$ will form the vertices of an Euclidean n-simplex Δ precisely when their images in $\mathbb{R}^{0,n}$ are $\mathbb{R}$-linearly independent. If the vertices are denoted by v_i, then we can define in a totally similar manner the exterior unit vectors u_i in $\mathbb{R}^n$. The points v of Δ are exactly the points of $\mathbb{R}^n$ satisfying the condition: $\langle v, u_i \rangle_1 \leq 0$ for each i. The corresponding Gram matrix $G(\Delta) = (\langle u_i, u_j \rangle_0)$ will have properties (13)–(17) and (18,0).

Given a matrix satisfying (13)–(17) and (18,0), it determines at most one Euclidean n-simplex Δ up to similarity (and central symmetry). In fact, there is a unique collection of signs $\epsilon_i = \pm$ (up to a uniform multiplication of all the ϵ_i by ± 1 which does not change the Gram matrix) so that the simultaneous column and row multiplication of the matrix will lead to the Gram matrix of Δ. To see this, we work in $\mathbb{V}(0)$ and note that the positivity of each of the $n \times n$ principal minors assures us that any subset of n of the vertices are $\mathbb{R}$-linearly independent. Thus, there is a unique linear relation (up to a non-zero scalar multiplier) of the form

$$\sum_i r_i u_i = 0, \ \epsilon_i r_i > 0, \ 0 \leq i \leq n, \ \epsilon_i = \pm 1. \tag{21}$$

Replacing u_i by $\epsilon_i u_i$ leads to the simultaneous multiplication of the i-th column and i-th row by ϵ_i. Up to a general linear transformation, it is enough to examine the special case where $u_i = -e_i$ for $1 \leq i \leq n$ and $u_0 = -(e_1 + \cdots + e_n)/n^{1/2}$. In this special case, we clearly have an n-simplex Δ, consisting of all vectors v in the form

$$v = e_0 + \sum_{1 \leq i \leq n} a_i e_i, \ a_i \geq 0, \text{ and } \sum_{1 \leq i \leq n} a_i \leq 1. \tag{22}$$

The 2^n different choices of signs ϵ_i would be applied to e_i and lead to 2^n different n-simplices obtained by appropriate hyperplane reflections. Together, they would form the generalized octahedron, or the regular crossed polytope dual to the n-dimensional cube, see [Coxeter, 1973]. For the general case, we may select the signs ϵ_i so that $r_i > 0$ for $i > 0$ and $r_0 < 0$ in (21). We then take any positive number N and define an Euclidean n-simplex associated to the Gram matrix by

$$\{v \in \mathbb{R}^n \mid \langle v, u_i \rangle \leq 0, 1 \leq i \leq n, \langle v, u_0 \rangle \leq N\}. \tag{23}$$

If N is replaced by $M > 0$, then the resulting n-simplex is similar to the preceding one under the transformation that sends v to $(N/M) \cdot v$.

For the hyperbolic n-space, there are several models at our disposal. We consider the $\mathbb{R}$-vector space $\mathbb{V}(-1)$ of dimension $n+1$ with inner product $\langle \cdot, \cdot \rangle_{-1}$; this is usually called the Minkowski space of signature $(1, n)$. We can then consider the set of all vectors v in $\mathbb{V}$ with $\langle v, v \rangle_{-1} = -1$. This set has two connected components. In terms of the basis e_i, each v can be written in the form

$$v = a_0 e_0 + \sum_{1 \leq i \leq n} a_i e_i, \quad -1 = -a_0^2 + \sum_{1 \leq i \leq n} a_i^2. \tag{24}$$

The two components are determined according to the sign of a_0. The usual convention is to take the component where $a_0 > 0$, namely, one of the two sheets of a hyperboloid. We note that each such vector v determines an open ray: $\mathbb{R}^+ v$. Using rays, we can then define the points at infinity. Each is an open ray consisting of vectors

$$\sum_{0 \leq i \leq n} a_i e_i, \quad a_0 > 0, \quad \text{and} \quad a_0^2 = \sum_{1 \leq i \leq n} a_i^2. \tag{25}$$

These rays then define the forward light cone of $\mathbb{V}(-1)$ with respect to $\langle \cdot, \cdot \rangle_{-1}$. We note that each of the rays defined so far has a unique intersection with the hyperplane defined by $a_0 = 1$. The collection of all the intersection points is the following closed n-disk:

$$(1, b_1, \cdots, b_n), \quad \sum_{1 \leq i \leq n} b_i^2 \leq 1. \tag{26}$$

As such, we have the Klein model of the hyperbolic n-space together with its ideal boundary at infinity. In this model, the geodesic subspaces are the intersections with the linear subspaces of $\mathbb{V}(-1)$. In particular, geodesics appear as chords joining two points on the boundary sphere at infinity. In terms of this negative-unit vector model, it is easy to check that the Gram matrix satisfies (13)–(17) and (18,–). Conversely, given an $(n+1) \times (n+1)$ matrix satisfying (13)–(17) and (18,–), there is a unique set of signs ϵ_i so that the modified matrix corresponds to a hyperbolic n-simplex with all of its vertices in the finite part of the hyperbolic n-space. To see this, we proceed as in the spherical case and define an abstract vector space over $\mathbb{R}$ of dimension $n+1$ with basis u_i and introduce the inner product using the given matrix as the Gram matrix. Sylvester's Law of Inertia together with (13)–(17) and (18,–) guarantee that our space is isomorphic to $\mathbb{V}(-1)$ with $\langle \cdot, \cdot \rangle_{-1}$. In other words, we may assume that u_i lies in $\mathbb{V}(-1)$. We next consider the linear subspace of $\mathbb{V}(-1)$ orthogonal to u_j for j not equal to i. Using (17) and the Theorem of Witt, we see that this linear subspace must be of dimension 1 and each nonzero vector v in this linear subspace must satisfy: $\langle v, v \rangle_{-1} < 0$. We may therefore define v_i so that $\mathbb{R}^+ v_i = \mathbb{R}^+ v$ lies inside the forward light cone so that $\langle v_i, v_i \rangle_{-1} = -1$. We now have $n+1$ points in the hyperbolic n-space. Thus, we have defined a unique hyperbolic n-simplex by taking the hyperbolic convex closure:

$$\Delta = \text{hccl}\{v_0, \ldots, v_n\} \qquad (27)$$

As things stand, we know that $\mathbb{R}^+ w$, $w = v_0 + \cdots + v_n$, is in the interior of Δ. It follows that $\langle u_i, w \rangle_{-1} \neq 0$. However, by choice, $\langle u_i, w \rangle_{-1} = \langle u_i, v_i \rangle_{-1}$. Thus, we must choose the signs $\epsilon_i = \pm 1$ so that $\langle \epsilon_i u_i, v_i \rangle_{-1} < 0$. The situation is now similar to the Euclidean case. In the Euclidean case, the forward light cone has opened up to a half space. In the spherical case, as indicated already, the correspondence between Gram matrices and congruence classes is unrestricted.

The extension to asymptotic simplices in hyperbolic n-space may now be described by the associated Gram matrices satisfying conditions (13)–(16), (18,−) and the following condition in place of (17):

(17,−)   all the $(n-1) \times (n-1)$ principal minors are positive definite,
   and all the $n \times n$ principal minors have non-negative determinant.

We note that conditions (14) and (17,−) force n to be greater than 1. When $n = 1$, an asymptotic simplex is either a closed ray or the entire hyperbolic line. In either case, it has infinite length. In contrast, when $n > 1$, an asymptotic hyperbolic n-simplex has finite n-dimensional volume.

Given an $(n+1) \times (n+1)$ matrix satisfying (13)–(16) and (17,−), (18,−), we proceed as before and form an abstract $\mathbb{R}$-vector space with basis u_i and inner product defined by the Gram matrix. Conditions (17,−) and (18,−) together with Sylvester's Law of Inertia assure us that the signature must be $(1, n)$. Equation (14) then assures us that the orthogonal complement to each u_i is an $\mathbb{R}$-subspace with signature $(1, n-1)$. Thus, its intersection with the forward ray model of hyperbolic n-space must be an $(n-1)$-dimensional hyperbolic subspace. As such, it is now easy to see that the $n \times n$ principal minors with determinant 0 will lead to a unique ray on the forward light cone, i.e., to a well-defined infinite point while the ones with positive determinant will again correspond to a finite point. Thus we obtain $n + 1$ points of the closed hyperbolic n-space (i.e., the hyperbolic n-space together with its ideal boundary.)

Each $n \times n$ principal minor with determinant zero in the Gram matrix corresponds to a vertex at infinity and defines an Euclidean simplex of dimension $n - 1$ up to similarity. This corresponds to the fact that the geometry of horospheres about a point at infinity of the hyperbolic n-space is Euclidean geometry up to similarity.

3 Scissors Congruence Groups

In the study of the Hilbert's Third Problem and its generalizations, it is convenient to introduce the scissors congruence group in all three classical geometries. With appropriate restrictions, these can be generalized. We will concentrate on the spherical and hyperbolic n-spaces: $\mathcal{P}(\mathbb{S}^n)$ and $\mathcal{P}(\mathcal{H}^n)$. We begin by forming the free abelian group generated by the symbols P, one for each n-dimensional geodesic polytope P. We then introduce the subgroups generated by the following elements:

If P and Q are interior disjoint with union R,

$$\text{then form } P + Q - R. \tag{28}$$

If σ is an isometry, and P is a geodesic polytope,

$$\text{then form } \sigma P - P. \tag{29}$$

The scissors congruence groups are defined to be the corresponding quotient groups. This definition originated with [Jessen, 1941]. It appears to be the first instance of a K-theory group. Evidently, the absolute volume map is a homomorphism from these groups onto the additive group of real numbers. The original Hilbert's Third Problem can be rephrased in the form: Show, by explicit examples, that the kernel is not zero when $n = 3$ and the geometry is Euclidean. This was definitively solved in [Dehn, 1901] for the Euclidean case.

Different versions of the problem had been considered by others. This caused some confusions in the literature. In the two non-Euclidean cases, Dehn only provided theoretical arguments rather than explicit examples. For all three geometries, it is known that the kernel is 0 when $n \leq 2$. When $n = 1$, it is essentially a tautology. In the case of Euclidean plane, the result was known to Euclid and completed in [Wallace, 1807] (cf. [Jackson, 1912]). For the 2-sphere, this follows from the spherical excess formula. For the hyperbolic plane, it follows from the defect formula of Gauss. This latter is best done by introducing the scissors congruence group for the extended hyperbolic n-space, $n > 1$, see [Sah, 1981].

In the definition of the scissors congruence group, the first task is to simplify the presentation. This led to quite a bit of controversies historically. For example, in [Bricard, 1896], a solution is sketched for a more restricted problem. At the time of the 1900 Paris Congress, it was not known that the restricted problem was equivalent to the one formulated by Hilbert. To be precise, the conditions (now called the Dehn invariants) formulated by Bricard were not obviously extendable to the formulation presented by Hilbert. To give the tensor product formulation of the Dehn invariant required us to show that (28) and (29) lead to the same scissors congruence group if we reduce the generating set to geodesic n-simplices and the division process to *simple bisections*. To be precise, a simple bisection of an n-simplex is obtained by the introduction of a vertex w along the edge joining two vertices. After that, we bisect the n-simplex using the geodesic hyperplane determined by w and the codimension 2 face opposite the edge containing w.

A proof of this was carried out in an unpublished Danish manuscript [Thorup, c. 1970]. For an algebraic topological proof of this result, see [Dupont, 1982]. Granting this result, we can now review the tensor product definition of the Dehn invariant due to [Jessen, 1968]. For our purposes, we will only describe it for the case of $n = 3$. For a general definition leading to Hopf algebra and comodule structures, see [Sah, 1979; Goncharov 1999]. Let Δ denote a 3-simplex. The Dehn invariant $D(\Delta)$ is an element of $\mathbb{R} \otimes \mathbb{R}/\mathbb{Z}\pi$ where the tensor product is over the ring of integers and the two factors are viewed as *discrete* abelian groups. In this setting, the definition is given by

$$D(\Delta) = \sum \ell(F) \otimes \theta(F), \tag{30}$$

F ranges over the edges of Δ; $\ell(F)$ is the length of the edge, and $\theta(F)$ is the interior dihedral angle along F in radians.

It is then easy to see that $D(\Delta)$ is additive with respect to simple bisection and is an isometry invariant. In this formulation, it is easy to see that the Euclidean unit cube and the regular tetrahedron of volume 1 have zero and nonzero Dehn invariant. Thus, they can not be scissors congruent. In passing, we should note that the scissors congruence group formulation automatically passes from scissors congruence to stable scissors congruence. However, under very general conditions which are met in the three classical cases, there is no difference between scissors congruence and stable scissors congruence, see [Zylev, 1965; Sah, 1979]. In order to bring in homological algebraic techniques, one has to show that the definition of the scissors congruence groups can be reformulated in the following manner, see [Dupont, 1982]. We fix an orientation of the n-dimensional space. For each ordered $(n + 1)$-tuples of points v_i, $0 \leq i \leq n$, lying in a small geodesic ball (this is a restriction only in the case of the n-sphere), consider a generator $(v_0, \ldots, v_n)$. We then introduce the following relations:

$$(v_0, \ldots, v_n) = 0 \text{ if all the } v_i\text{'s lie in a geodesic } (n - 1)\text{-subspace}, \tag{31}$$

$$\sum_i (-1)^i (v_0, \ldots, v_{i-1}, v_{i+1}, \ldots, v_{n+1}) = 0 \text{ for any } (v_0, \ldots, v_{n+1}), \quad (32)$$

$$(g(v_0), \ldots, g(v_n)) = \det(g) \cdot (v_0, \ldots, v_n), g \text{ denotes an isometry. } (33)$$

The orientation is then used to identify this scissors congruence group with the scissors congruence groups defined earlier. It should be noted that extensions are possible for extended hyperbolic n-space where the vertices are allowed to be infinite, cf. [Dupont–Sah, 1982]. However, one has to be very careful. In the case of $n = 3$, there is no problem. We can in fact assume that all the vertices are at infinity or we can assume that all the vertices are in the finite part of the hyperbolic 3-space. This difficulty already makes an appearance in the case of the hyperbolic plane. Here, there is just one totally asymptotic triangle with area π. The actual scissors congruence group is isomorphic to $\mathbb{R}$ under the area map. In particular, there is a distinction between even and odd dimension. This is connected with the Gauss-Bonnet Theorem, see [Sah, 1981]. We now summarize the connection of the scissors congruence groups with various integral homology groups of the associated groups of isometries viewed as *discrete* groups.

The basic result is the following exact sequence, see [Dupont–Sah, 1982], which is a modification of an exact sequence first considered by Bloch–Wigner in [Bloch, 1978], see also [Suslin, 1991].

$$0 \to \mathbb{Q}/\mathbb{Z} \to H_3(SL(2, \mathbb{C})) \to \mathcal{P}(\mathbb{C}) \xrightarrow{\lambda} \Lambda_{\mathbb{Z}}^2(\mathbb{C}^\times)$$
$$\to H_2(SL(2, \mathbb{C})) \to 0 \qquad (34)$$

This exact sequence is the result of a detailed examination of an equivariant (also called a transposed) spectral sequence arising from the action of $SL(2, \mathbb{C})$ on the projective line $P^1(\mathbb{C})$ which is identified with the ideal boundary of the hyperbolic 3-space. In fact, the analysis can be carried out for a general division ring $\mathbb{D}$. However, the corresponding spectral sequence would lead to changes in the exact sequence (34).

In general, compare [Dupont–Parry–Sah, 1988], for any division algebra $\mathbb{D}$, we define the scissors congruence group $\mathcal{P}_{\mathbb{D}}$ as an abelian group with generators $[x]$, $x \in \mathbb{D} - \{0, 1\}$. These are called the cross-ratio symbols and correspond to ordered set of 4-tuples: $(\infty, 0, 1, x)$ of elements of the projective line $P^1(\mathbb{D})$. We then impose the defining relations

$$[x] = [yxy^{-1}] \text{ for } x, y \text{ in } \mathbb{D} - \{0, 1\}; \qquad (35)$$

$$[x] - [y] + [x^{-1}y] - [(x-1)^{-1}(y-1)] + [(x^{-1}-1)^{-1}(y^{-1}-1)] = 0, \quad (36)$$
where $x \neq y \in \mathbb{D} - \{0, 1\}$.

A second scissors congruence group $\mathcal{P}(\mathbb{D})$ is defined by using (35), (36) together with the following additional relations:

$$[x] + [x^{-1}] = 0, \ x \in \mathbb{D} - \{0, 1\}; \tag{37}$$

$$[x] + [1 - x] = constant \ (\text{depending on } \mathbb{D}), \ x \in \mathbb{D} - \{0, 1\}. \tag{38}$$

In general, we have [Sah, 1989, (2.11)]: If $|\mathbb{D}| \geq 5$, then there is an exact sequence

$$\mathbb{D}^{\times}/(\mathbb{D}^{\times})^2 \to \mathcal{P}_{\mathbb{D}} \to \mathcal{P}(\mathbb{D}) \to 0, \tag{39}$$

where $(\mathbb{D}^{\times})^2$ is the subgroup generated by the squares.

For the three classical real division algebras: $\mathbb{R}$, $\mathbb{C}$, $\mathbb{H}$, the first map in (39) is injective. Thus, there is no difference between $\mathcal{P}(\mathbb{D})$ and $\mathcal{P}_{\mathbb{D}}$ when $\mathbb{D} = \mathbb{C}$ or $\mathbb{H}$ and $\mathcal{P}_{\mathbb{R}}$ maps onto $\mathcal{P}(\mathbb{R})$ with kernel of order 2. The constant in (38) is 0 for $\mathbb{D} = \mathbb{C}$ or $\mathbb{H}$ and has order 3 when $\mathbb{D} = \mathbb{R}$.

In the exact sequence (34), the map λ carries $[x]$ onto $x \wedge (1 - x)$ in the abelian group $\Lambda_{\mathbb{Z}}^2(\mathbb{C}^{\times})$. $H_2(SL(2, \mathbb{C}))$ can be identified with $K_2(\mathbb{C})$, the algebraic K_2-group of the field $\mathbb{C}$ of complex numbers. This uses a homological stability theorem, see [Suslin, 1984; Sah, 1986], together with the 2-divisibility of the multiplicative group of complex numbers. Since $\mathbb{C}$ is algebraically closed, it follows from [Bass–Tate, 1973] that $K_2(\mathbb{C})$ is a $\mathbb{Q}$-vector space of dimension equal to the transcendence degree of $\mathbb{C}$ over $\mathbb{Q}$. The group $\mathcal{P}(\mathbb{C})$ is known to be a $\mathbb{Q}$-vector space. The divisibility was proved in [Dupont–Sah, 1982]. This is a corollary of the distribution formula

$$[x^n] = n \cdot \sum_{0 \leq j \leq n-1} [\zeta^j x], \ \zeta \text{ a primitive } n\text{-th root of 1.} \tag{40}$$

The absence of torsion depends on Suslin's affirmative resolution of the Lichtenbaum–Quillen conjecture on the algebraic K-theory of algebraically closed fields F [Suslin, 1986, 1991]. Thus, for $n > 0$, $K_{2n}(F)$ is a $\mathbb{Q}$-vector space, $K_{2n-1}(F)$ is the direct sum of a $\mathbb{Q}$-vector space together with the group of roots of 1 in F. Using homological stability, it is then known that $H_3(SL(2, F))$ is isomorphic to the indecomposable part of $K_3(F)$, see [Suslin, 1991; Sah, 1989]. For $F = \mathbb{C}$, this yields the result that $\mathcal{P}(\mathbb{C})$ is a $\mathbb{Q}$-vector space.

To connect up (34) with the scissors congruence problems in hyperbolic and spherical 3-space, we need to examine the action of complex conjugation automorphism of $\mathbb{C}$. More generally, we can examine the action of the Galois automorphisms of $\mathbb{C}$ over $\mathbb{Q}$. The complex conjugation automorphism of $\mathbb{C}$ induces the orientation reversing isometry on hyperbolic

3-space. The group $\mathbb{Q}/\mathbb{Z}$ is $\mathrm{Tor}_1(\mathbb{C}^\times, \mathbb{C}^\times)$ and is pointwise fixed by complex conjugation. The hyperbolic scissors congruence group in dimension 3 is then related to the negative eigen-space of the complex conjugation action on (34). This led to the exact sequence

$$0 \to H_3(SL(2,\mathbb{C}))^- \to \mathcal{P}(\mathbb{C})^- = \mathcal{P}(\mathcal{H}^3)$$
$$\xrightarrow{D} \mathbb{R} \otimes (\mathbb{R}/\mathbb{Z}\pi) \to H_2(SL(2,\mathbb{C}))^- \to 0. \quad (41)$$

Strictly speaking, we needed to show that $\mathcal{P}(\mathcal{H}^3) = \mathcal{P}(\partial \mathcal{H}^3)$, i.e., the scissors congruence group in hyperbolic 3-space can be identified with the scissors congruence group generated by the totally asymptotic hyperbolic 3-simplices. This depends on 3 being odd. The classical hyperbolic Dehn invariant map is then the negative part of λ with respect to complex conjugation. The positive eigen-spaces of the exact sequence (34) is related to the scissors congruence problem in spherical 3-space in a more complex manner. To be precise, we map the following exact sequences (42), (43) into the exact sequence (44) consisting of the space of co-invariants in (34) under the action of the complex conjugation, see [Dupont–Parry–Sah, 1988, Section 5]:

$$0 \to H_3(SU(2)) \to \mathcal{P}(\mathbb{S}^3)/\mathbb{Z} \to \mathbb{R} \otimes (\mathbb{R}/\mathbb{Z}) \to H_2(SU(2)) \to 0. \quad (42)$$
$$0 \to \mathbb{Q}/\mathbb{Z} \to H_3(SU(2)) \to \mathcal{P}(\mathbb{S}^3)/(\mathcal{P}(\mathbb{S}^1) \# \mathcal{P}(\mathbb{S}^1))$$
$$\to \Lambda_{\mathbb{Z}}^2(\mathbb{R}/\mathbb{Z}) \to H_2(SU(2)) \to 0. \quad (43)$$

$$0 \to \mathbb{Q}/\mathbb{Z} \to H_0(\langle * \rangle, H_3(SL(2,\mathbb{C}))) \to \mathcal{P}(\mathbb{C}, *)$$
$$\xrightarrow{\lambda} \Lambda_{\mathbb{Z}}^2(\mathbb{R}^+) \oplus \Lambda_{\mathbb{Z}}^2(\mathbb{R}/\mathbb{Z}) \to H_2(SL(2,\mathbb{C}))^+ \to 0. \quad (44)$$

It should be noted that the torsion subgroup $\mathbb{Q}/\mathbb{Z}$ of $H_3(SU(2))$ corresponds exactly to the subgroup of $\mathcal{P}(\mathbb{S}^3)/\mathbb{Z}$ consisting of the rational lunes. The isomorphism $H_2(SU(2)) \cong H_2(SL(2,\mathbb{C}))$ was proved in [Dupont–Parry–Sah, 1988] improving the surjectivity result of [Mather, 1975; Sah–Wagoner, 1977]. At the same time, in a slightly different form, the surjectivity from $H_3(SU(2))$ to $H_0(\langle * \rangle, H_3(SL(2,\mathbb{C})))$, was proved and conjectured to be an isomorphism. The injectivity is then proved in [Bökstedt–Brun–Dupont, 1998], cf. [Dupont–Parry–Sah, 1988, Remark 4.8]. The main task is to show that any hyperbolic n-simplex is homologous to a sum and difference of hyperbolic n-simplices that can be circumscribed on a "finite sphere" in a systematic way. Thus, we have:

$$H_3(SU(2)) \cong H_0(\langle * \rangle, H_3(SL(2,\mathbb{C}))) = H_3(SL(2,\mathbb{C}))^+. \quad (45)$$

In particular, we have the exact sequence

$$0 \to \mathcal{P}(\mathbb{S}^3)/(\mathcal{P}(\mathbb{S}^1)\#\mathcal{P}(\mathbb{S}^1)) \to \mathcal{P}(\mathbb{C},*) \to \mathcal{P}(\mathbb{H}) = \Lambda_{\mathbb{Z}}^2(\mathbb{R}^+) \to 0. \quad (46)$$

These exact sequences then exhibit the Galois action on (34) and the relation with the scissors congruence problems in hyperbolic and spherical 3-space. However, it is not easy to see the map in (46) in a transparent way. In particular, to map a spherical 3-simplex into $\mathcal{P}(\mathbb{C},*)$, it is necessary to use the Hopf fibration. It is also important to note that a general Galois automorphism of $\mathbb{C}$ over $\mathbb{Q}$ does not commute with the familiar complex conjugation map. In fact, it is a known result that the only Galois automorphism of an algebraically closed field of finite order must be of order 2 and its fixed field must be real closed and the algebraic closure is then obtained by adjoining the square root of -1. Thus, all the order 2 automorphisms of $\mathbb{C}$ are conjugate to complex conjugation in the full group of automorphisms of $\mathbb{C}$. For each such automorphism, we would have two corresponding exact sequences related to the scissors congruence problem in dimension 3 spherical and hyperbolic spaces. In order to see the Galois action, we note that $PSL(2,\mathbb{C}) \cong SO^1(1,3;\mathbb{R})$ and the action of $PSL(2,\mathbb{C})$ on $P^1(\mathbb{C})$ is compatible with the action of $SO^1(1,3;\mathbb{R})$ on the ideal boundary of hyperbolic 3-space modelled in $\mathbb{V}(-1)$. At the same time, $SO(4;\mathbb{R})$ acts on the unit 3-sphere in $\mathbb{V}(1)$. If we complexify both $\mathbb{R}$-vector spaces, we can then consider $SO^1(1,3;\mathbb{R})$ and $SO(4;\mathbb{R})$ as real forms of $SO(4,\mathbb{C})$, or subgroups of $SL(4,\mathbb{C})$. We may then invoke homological stability theorems for H_i, $i \leq 3$. However, the complexification map involves a different embedding and leads to a multiplication by 2 isogeny of the relevant algebraic K-groups. At the level of $H_3(SL(2,\mathbb{C}))$, there will be a kernel of order 2. Since we know the structure of these homology groups as well as the source of the element of order 2, there is no serious problem. There still remains the question about what exactly happened to the scissors congruence groups. In [Dupont–Parry–Sah, 1988], the group $\mathcal{P}(\mathbb{S}^3)/[\text{point}]\#\mathcal{P}(\mathbb{S}^2)$ was identified with the group

$$H_0(\langle *\rangle, H_3(C_*(\mathbb{S}^3) \otimes_{SO(4;\mathbb{R})} \mathbb{Z}^t)),$$

where $\mathbb{Z}^t$ denotes the twisted action.

This went by way of the map sending a simplex to the corresponding oriented tuple minus its mirror image. Under complexification, the boundary sphere is replaced by $P^1(\mathbb{C}) \times P^1(\mathbb{C})$ with $SO(4,\mathbb{C}) \cong SL(2,\mathbb{C}) \times SL(2,\mathbb{C})/\langle -I_4\rangle$ acting on each factor. By Künneth's Theorem, the homology group is just two copies of $\mathcal{P}(\mathbb{C})$. Since $\langle *\rangle$ interchanges the two factors, we just have one copy of $\mathcal{P}(\mathbb{C})$ left. It is in this group where the comparison of spherical and dual hyperbolic simplices take place.

4 Galois action

In this section, we examine the action of the full Galois group $\mathcal{G} = \mathrm{Gal}(\mathbb{C}/\mathbb{Q})$ of field automorphisms of the complex numbers on the exact sequence (34) and relate it to the geometry of Gram matrices. $\mathcal{G}$ also acts on the $\mathbb{Q}$-vector space of all $(n+1) \times (n+1)$ symmetric complex matrices. Although (14) is unchanged, the reality condition (13) is usually violated. Even if (13) is retained for a specific automorphism and a specific Gram matrix $G(\Delta)$, there is no assurance that (17) is unchanged. In the optimal situation where $G(\Delta)$ and $\sigma(G(\Delta))$ are both Gram matrices, it is totally possible that they are of different types. Since the determinant of a matrix is a polynomial with integer coefficients in its entries, it is immediate that $G(\Delta)$ is Euclidean (respectively k-asympotic) if and only if $\sigma(G(\Delta))$ is Euclidean (respectively k-asymptotic).

More interestingly, there is the possibility that one of $G(\Delta)$, $\sigma(G(\Delta))$ is spherical while the other is hyperbolic. In each of these cases, it may be necessary to extract some square roots before we can meaningfully define the action of σ on the vertices of Δ. The ambiguity amounts to sign changes on the Gram matrices. Evidently, σ is compatible with sign changes. As noted in section 2, except in the case of a Gram matrix of spherical type, each of the other types of Gram matrix determines a unique n-simplex (up to isometry or similarity) after appropriate change of signs. When this is the case, we can speak of a Galois conjugate of an n-simplex Δ. The sign change is not necessary when $\sigma(G(\Delta))$ is spherical. However, if $G(\Delta)$ is spherical and $\sigma(G(\Delta))$ is hyperbolic, then it may still be necessary to change some of the signs of $G(\Delta)$. As noted in (19), sign changes in $G(\Delta)$ can only change the scissors congruence class of Δ modulo lunes by a sign of ± 1. It should be noted that the Galois action does not usually extend to the interior of n-simplices. When it does, it amounts to a permutation of the vertices.

The preceding discussion applies when $\mathbb{C}$ is replaced by the algebraic closure $\mathbb{C}_{alg}$ of $\mathbb{Q}$. It is especially interesting when $\mathbb{C}$ is replaced by $\mathbb{Q}^{ab}$, the maximal abelian extension of $\mathbb{Q}$. By the Theorem of Kronecker–Weber, $\mathbb{Q}^{ab}$ is generated by all the roots of 1. If $\zeta = \exp(2\pi \iota j/m)$ is an m-th root of 1, then $\cos 2\pi \iota j/m = (\zeta + \zeta^{-1})/2$ is totally real. In particular, it is fixed under complex conjugation. For any Galois automorphism σ of $\mathbb{C}$, $\sigma(\zeta)$ is another m-th root of 1. In general, an n-simplex Δ is called *rational* if all its dihedral angles $\theta_{i,j}$ are in $\mathbb{Q}\pi$. It follows that a Galois conjugate of a rational n-simplex (when it is defined) is again a rational n-simplex.

From now on, we concentrate our attention to the case $n = 3$. We note:

A rational 3-simplex always has Dehn invariant 0. (47)

Lemma 4 *Let P be a spherical polytope of dimension 3. Then,*

(a) *If P is a lune, then equivalent statements are: (1) P has zero Dehn invariant (2) P is scissors congruent to a rational lune; and (3) P has rational volume.*

(b) *Let P be a 3-simplex with P' denoting one of the 2^4 simplices obtained from sign changes of its vertices. Then P is rational if and only if P' is rational.*

Proof. (a) We note that a lune in spherical 3-space is the orthogonal suspension of a spherical triangle with vertex angles α, β, γ. The latter is scissors congruent to the orthogonal suspension of an arc, see [Sah, 1981]. Thus, a lune in spherical 3-space is scissors congruent to the orthogonal join of an arc of length $\pi/2$ with an arc of length δ. Its Dehn invariant is therefore $\pi/2 \otimes \delta \in \mathbb{R} \otimes (\mathbb{R}/\mathbb{Z}\pi)$. By the proportionality principle, its volume is $\pi\delta/4$. Thus, (a) holds. (b) follows from (a) and (19). $\square$

Remark Using Lemma 4, it is not very difficult to check and see that the rational spherical simplices considered in [Coxeter, 1935] are all scissors congruent to lunes. The starting point in each case was a fundamental domain of a finite Coxeter group acting on the 3-sphere. These then have rational volumes and are scissors congruent to lunes as shown in [Dupont–Sah, 1982, Theorem 6.4]. The rest of the cases then follows by using (b) of Lemma 4. In essence, we have bypassed the analytic arguments used by Coxeter and Schläfli and replaced them by cutting and pasting.

For the remaining part of this section, we will concentrate our attention on orthoschemes, see [Schläfli, 1860; Coxeter, 1935]. The Gram matrices have the following form:

$$G(\alpha, \beta, \gamma) = \begin{bmatrix} 1 & -a & 0 & 0 \\ -a & 1 & -b & 0 \\ 0 & -b & 1 & -c \\ 0 & 0 & -c & 1 \end{bmatrix}, \tag{48}$$

where $a = \cos\alpha$, $b = \cos\beta$, $c = \cos\gamma$, $0 < \alpha, \gamma < \pi/2$, $0 < \beta < \pi$.

When $G(\alpha, \beta, \gamma)$ corresponds to a geometric simplex, this 3-simplex will be denoted by $\Delta(\alpha, \beta, \gamma)$, it then corresponds to $[\pi/2 - \alpha, \beta, \pi/2 - \gamma]$ in [Coxeter, 1935]. α, β, γ then correspond to 3 of the interior dihedral angles while the remaining three are $\pi/2$. The matrix in (48) satisfies (17) if and only if:

$$\cos^2 \beta < \sin^2 \alpha, \ \sin^2 \gamma, \quad \text{that is} \quad 0 < \alpha, \ \gamma < \pi/2 < \alpha + \beta, \ \beta + \gamma. \quad (49)$$

We note that $\det(G(\alpha,\beta,\gamma)) = (1-a^2)(1-c^2)-b^2 = \sin^2 \alpha \cdot \sin^2 \gamma - \cos^2 \beta$. Thus, $G(\alpha,\beta,\gamma)$ is the Gram matrix of a spherical orthoscheme if and only if

$$\cos^2 \beta < \sin^2 \alpha \cdot \sin^2 \gamma. \quad (50,+)$$

The hyperbolic case is more complicated. $G(\alpha,\beta,\gamma)$ corresponds to a hyperbolic orthoscheme if and only if

$$\sin^2 \alpha \cdot \sin^2 \gamma < \cos^2 \beta < \sin^2 \alpha, \ \sin^2 \gamma. \quad (50,-)$$

If the inequalites on the right of (50,−) are replaced by equalities, then we obtain an asymptotic orthoscheme. They are either 1- or 2-asymptotic depending on the number of equalities. Condition (50,−) is insensitive to sign changes in (49). However, as mentioned before, there is a unique choice that leads to well-defined hyperbolic 3-simplex $\Delta(\alpha,\beta,\gamma)$. The values of these angles are sensitive to the sign changes.

Example Consider $G(\alpha,\beta,\gamma)$ with $\alpha = \pi/5$, $\beta = \pi/3$, and $\gamma = \pi/5$. Since we have

$$\cos^4(3\pi/10) < \cos^2(\pi/3) < \cos^2(3\pi/10),$$

$G(\pi/5,\pi/3,\pi/5)$ corresponds to a hyperbolic 3-simplex. In fact, this simplex is the fundamental domain for the hyperbolic Coxeter group corresponding to the diagram, cf. Bourbaki, Groupes et algèbres de Lie, Chapter V, p. 133,

$$\bullet \overset{5}{\rule{2em}{0.4pt}} \bullet \overset{3}{\rule{2em}{0.4pt}} \bullet \overset{5}{\rule{2em}{0.4pt}} \bullet$$

It has a unique Galois conjugate $G(3\pi/5,\pi/3,3\pi/5)$. Since we have

$$\cos(\pi/3) < \cos^2(\pi/5) = (3 + 5^{1/2})/8,$$

the Galois conjugate $G(3\pi/5,\pi/3,3\pi/5)$ is a rational spherical 3-simplex. The following lemma provides infinitely many examples of this kind.

Lemma 5 *Let $G(\alpha,\beta,\gamma)$ be a Gram matrix of a rational orthoscheme Δ of spherical or hyperbolic type. Let $\exp(\iota\alpha)$, $\exp(\iota\beta)$ and $\exp(\iota\gamma)$ have order 2p, 2q, and 2r respectively as roots of 1. Then,*

(a) *For any automorphism σ of $\mathbb{C}$, $\sigma(G(\alpha, \beta, \gamma))$ is either of spherical type or hyperbolic type.*

(b) *If $q = 2$, equivalently, $\beta = \pi/2$, then $G(\alpha, \beta, \gamma)$ corresponds to a spherical simplex which is the orthgonal join of two rational arcs; it is then scissors congruent to a lune and has rational volume.*

(c) *Suppose $\alpha = \gamma$, thus $p = r$, and suppose p, q are distinct prime numbers with $1/6 > 4/q + 1/p$. Then there exist Galois automorphisms σ and τ of $\mathbb{C}$ so that the following hold:*

(c₁) *$\sigma(G(\alpha, \beta, \gamma))$ determines a spherical 3-simplex $\Delta(\sigma)$ and $\tau(G(\alpha, \beta, \gamma))$ determines a hyperbolic 3-simplex $\Delta(\tau)$.*

(c₂) *$\Delta(\sigma)$ is not scissors congruent to a lune.*

In fact, the argument we present applies to any Gram matrix and any Galois automorphism fixing all but one row and the corresponding column.

Proof.

(a) Two of the three 3×3 principal minors of $\sigma(G(\alpha, \beta, \gamma))$ are positive definite. Since $\det(\sigma(G(\alpha, \beta, \gamma))) = \sigma(\det(G(\alpha, \beta, \gamma))) \neq 0$, (a) holds.

(b) $G(\alpha, \beta, \gamma)$ is the direct sum of two 2×2 blocks. The assertion is then straight forward.

(c₁) Since p and q are distinct odd primes, the corresponding cyclotomic field of $2p$-th and $2q$-th roots of 1 are linearly disjoint. Thus, we can find Galois automorphisms carrying $\exp(\iota\alpha)$ and $\exp(\iota\beta)$ independently to arbitrary $2p$-th and $2q$-th roots of 1.

In order to find σ, we just have to choose α and β to satisfy the following inequality:

$$\cos^2 \beta < \sin^4 \alpha; \text{ equivalently, } 0 < |\cos \beta| < \sin^2 \alpha.$$

Since $q > 3$, we can find β so that $\pi/3 < \beta < 2\pi/3$, thus, $0 < |\cos \beta| < 1/2$. Similarly, since $p > 5$, we can find α so that $\pi/4 < \alpha < 3\pi/4$, thus, we have: $1/2 < \sin^2 \alpha$.

In order to find τ, we must find α and β to satisfy the following inequalities:

$$\sin^4 \alpha < \cos^2 \beta < \sin^2 \alpha.$$

We begin by taking $0 < \alpha < \pi/6$ so that $0 < \sin \alpha < 1/2$. We then have

$$\sin(\alpha/2) - \sin^2 \alpha = \sin(\alpha/2) \cdot [1 - 2 \cdot \cos(\alpha/2) \cdot \sin \alpha] > 0.$$

In other words, it is enough to find α and β so as to satisfy:

$$0 < \alpha/2 < |\pi/2 - \beta| < \alpha < \pi/6.$$

We note our assumption guarantees $p/6 - 4p/q > 1$. Thus, we can find an integer t so that $4p/q < t < p/6$. We take $\alpha = t\pi/p$ so that $\alpha < \pi/6$ holds. Between α and $\alpha/2$, we have an angular interval $\alpha/2$ or $t\pi/2p$ which, by the above inquality, is greater than $2\pi/q$. Since the primitive $2q$-th roots of 1 together with -1 are uniformly distributed around the unit circle with successive angular gaps of $2\pi/q$ radians, we may invoke Dirichlet's Box Principle to find β so that $|\pi/2 - \beta|$ falls strictly between $\alpha/2$ and α. The strictness is a consequence of the fact that p and q are distinct primes. We note also that β could not be equal to π so that it must correspond to a primitive $2q$-th root of 1.

(c_2). $\Delta(\sigma)$ has zero Dehn invariant. Thus, according to (42), it determines a homology class $h(\Delta(\sigma))$. If $\Delta(\sigma)$ were scissors congruent to a lune, then this class must be of finite order. Since $H_3(SU(2))$ is mapped into $H_3(SL(2, \mathbb{C}))$, we can first apply the Galois automorphism $\tau \circ \sigma^{-1}$ to $h(\Delta(\sigma))$ and follow this by the imaginary part of the second Cheeger–Chern–Simons homomorphism $\hat{C}_2$, see [Dupont, 1987]. This latter yields the hyperbolic volume of $\Delta(\tau)$ which is non-zero. This shows that $h(\Delta(\sigma))$ has infinite order. Thus, $\Delta(\sigma)$ can not be scissors congruent to a lune. $\square$

Remarks In (c), the restriction on p and q automatically holds when $p, q > 30$.

In the proof of (c_2), it is necessary to check the compatibility of the maps. In particular, the switch from the spherical to the hyperbolic set up seems mysterious. As noted at the end of section 3, the compatibility is implicit in [Dupont–Parry–Sah, 1988].

More generally, it is possible to consider the case where $\tau(G(\alpha, \beta, \gamma))$ is of hyperbolic type whether it corresponds to a hyperbolic simplex or not. Here, one or two vertices in the dual simplex may be ultra-infinite. This means that they correspond to vectors v in $\mathbb{V}(-1)$ with $\langle v, v \rangle_{-1} = 1$. In such a case, we cut away the infinite part of the dual simplex by using the hyperbolic plane orthogonal to v_0. This corresponds to the construction of the orthogonal cone in the spherical space. In essence, we look at $G(0, \beta, \gamma)$ or $G(\alpha, \beta, 0)$. By a continuity argument, it is not difficult to see that the cutting plane must be on one side or the other of the face opposite the ultra-infinite vertex. Thus, we are left with an honest compact hyperbolic polyhedron with non-zero volume. In [Bökstedt–Brun–Dupont, 1998], the homological stability theorem was proved using a chain complex that involved both finite as well as ultra-infinite vertices. This provides the justification for the preceding truncation process. Since the modification involves adding or subtracting off a rational lune of spherical type associated to $G(0, \beta, \gamma)$ and/or $G(\alpha, \beta, 0)$, we know that we have only added to $h(\Delta(\sigma))$ an element of finite order in $H_3(SU(2))$. Thus, the preceding lemma can be strengthened to:

Theorem 6 *Let $G(\alpha, \beta, \gamma)$ be the Gram matrix of a rational spherical orthoscheme Δ. If there is a Galois automorphism τ of $\mathbb{C}$ such that $\tau(G(\alpha, \beta, \gamma))$ is of hyperbolic type, then the element $h(\Delta)$ of $H_3(SU(2))$ determined by Δ must have infinite order. Conversely, if every $\tau(G(\alpha, \beta, \gamma))$ is of spherical type, i.e., with positive determinant, then Δ is scissors congruent to a lune, and has rational volume.*

Proof. We only have to justify the converse. This uses the result of [Bökstedt–Brun–Dupont, 1998] showing that the map $H_3(SU(2)) \rightarrow H_3(SL(2, \mathbb{C}))$ is injective. We next note that the vertices of Δ are algebraic so that the class lies in $H_3(SL(2, \mathbb{C}_{alg}))$. It follows from the theorem of Borel on regulators [Borel, Collected Papers, 108, 1977], that the non-torsion elements are detected by using the collection of all the imaginary parts of the Galois twists of the Cheeger–Chern–Simons class. Since all the Galois conjugates are spherical, our class must be a torsion class in $H_3(SL(2, \mathbb{C}))$. Since $H_3(SU(2))$ captures all the torsion elements, we conclude that $h(\Delta)$ is torsion. But the torsion classes are represented by lunes. Thus, from [Bökstedt–Brun–Dupont, 1998], we conclude that Δ is scissors congruent to a lune. □

Proof of Theorem 2. In fact, the first statement is a reformulation of Lemma 4 by comparing Δ in (b) with the lune of the same volume. The second statement is just Theorem 6 and the last follows from the examples given in Lemma 5 (b). □

We now go on to Theorem 3. This time, we will be dealing with totally asymptotic simplices. The Galois conjugate of such a simplex need not be hyperbolic, but when it does it remains totally asymptotic. In fact, the Galois action is directly described in terms of the action on $\mathcal{P}(\mathbb{C})$. It is not difficult to see, cf. [Dupont–Sah, 1982], that the totally asymptotic 3-simplex $(\infty, 0, 1, z)$ with $z = \exp(2\iota\theta)$ is scissors congruent to the sum

of 4 copies of
$$T(\theta) = \Delta(\theta, \pi/2 - \theta, \theta), \; 0 < 2\theta < \pi.$$
Its volume is $\text{Л}(\theta)$ and its Dehn invariant $D(\infty, 0, 1, z) = 2\log(2\sin\theta) \otimes \theta$. In particular we note, equivalent statements are

$$\lambda[z] = 0, \qquad\qquad\qquad (51, a)$$

$$D(\infty, 0, 1, z) = 0, \qquad\qquad\qquad (51, b)$$

$$z = \zeta_n \text{ is an } n\text{-th root of unity for some } n. \qquad (51, c)$$

Furthermore, the following was proved by [Bloch, 1978], (cf. also [Gross, c. 1980] and [Parry, c. 1984], all unpublished).

Theorem 7 *The subgroup of $\mathcal{P}(\mathbb{C})$ generated by $\{[z] \mid z \in \mathbb{Q}[\zeta_n]\}$ equals $K_3(\mathbb{Q}[\zeta_n])/\text{torsion}$, and has $\mathbb{Q}$-basis $\{[\zeta_n{}^k] \mid 0 < k < n/2, \; \gcd(k, n) = 1\}$.*

Remark It follows that this subgroup is contained in $\mathcal{P}(\mathbb{C})^- = \mathcal{P}(\partial\mathcal{H}^3)$ and thus Milnor's conjecture is valid at the level of the scissors congruence group. Hence, by (51), Milnor's conjecture is a special case of the Generalized Hilbert's Third Problem.

Proof of Theorem 3. Given the form of the graph of Л on $[0, \pi/2]$, we have a real analytic map $t : [0, \pi/2] \to [0, \pi/2]$ of order 2 with fixed point $\pi/6$ so that $\text{Л}(\theta) = \text{Л}(t(\theta))$

Let θ range over the rational multiples of π in $[0, \pi/2]$. We now have two cases:

(a) If $t(\theta)$ is irrational, then $T(\theta)$ has zero Dehn invariant while $T(t(\theta))$ has non-zero Dehn invariant so that they furnish explicit counter examples for the classical Hilbert Third Problem for hyperbolic 3-space as desired by Dehn.

(b) If $t(\theta)$ is rational, then Milnor's conjecture is false. In fact we can assume

$$\theta = \pi \cdot (p/n), t(\theta) = \pi \cdot (q/n), \text{ with } p, q < n/2, \; \gcd(p, n) = 1.$$

Now, if also $\gcd(q, n) = 1$, then by Theorem 4.9, we get a contradiction to Milnor's conjecture. On the other hand, if $\gcd(q, n) > 1$, then there is a Galois automorphism of $\mathbb{Q}[\zeta_n]$ which fixes $[\zeta_n{}^q]$ but moves $[\zeta_n{}^p]$ in $K_3(\mathbb{Q}[\zeta_n]) \otimes \mathbb{Q}$. In particular,

$$[\zeta_n{}^p] \neq [\zeta_n{}^q],$$

this again contradicts Milnor's conjecture since by construction, $\text{Л}(\theta) = \text{Л}(t(\theta))$. $\square$

References

H. Bass and J. Tate, *The Milnor Ring of a Global Field,* Springer LNM 342 (1973), 349–446.

S. J. Bloch, *Higher Regulators, Algebraic K-theory, and Zeta Functions of Elliptic Curves,* Irvine Lecture Notes, 1978, unpublished.

M. Bökstedt, M. Brun, and J. L. Dupont, Homology of $O(n)$ and $O^1(1,n)$ made discrete: an application of edgewise subdivision, *J. Pure Appl. Algebra* **123** (1998), 131–152.

A. Borel, Cohomologie de SL_n et valeurs de fonctions zeta aux points entiers, *Ann. Sci. Norm. Sup. Pisa, Cl. Sci.* **4** (1977), 613–616, correction, ibid. **7** (1980) 373. See also Collected Papers, 3 (1983), 495–519.

R. Bricard, Sur une question de géométrie relative aux polyèdres, *Nouv. Ann. Math.* **15** (1896), 331–334.

J. L. Cathelineau, Homology of tangent groups considered as discrete groups and scissors congruence, *J. Pure Appl. Algebra* **132** (1998), 9–25.

J. Cheeger and J. Simons, *Differential Characters and Geometric Invariants,* Springer LNM 1167 (1985), 50–80.

H. S. M. Coxeter, The functions of Schläfli and Lobatschefsky, *Quart. J. Math.* **6** (1935), 13–29.

H. S. M. Coxeter, *Regular Polytopes,* 3rd ed., Dover, 1973.

M. Dehn, Über den Rauminhalt, *Math. Ann.* **105** (1901), 465–478.

M. Dehn, Die Eulersche Formel im Zussammenhang mit dem Inhalt in der Nicht-Euklidischen Geometrie, *Math. Ann.* **61** (1906), 561–586.

J. L. Dupont, Algebra of polytopes and homology of flag complexes, *Osaka J. Math.* **19** (1982), 599–641.

J. L. Dupont, The dilogarithm as a characteristic class for flat bundles, *J. Pure Appl. Algebra* **44** (1987), 137–164.

J. L. Dupont, W. Parry and C. H. Sah, Homology of classical Lie groups made discrete, II, *J. Algebra* **113** (1988), 215–260.

J. L. Dupont and C. H. Sah, Scissors congruences, II, *J. Pure and Appl. Algebra* **25** (1982), 159–195.

J. L. Dupont and C. H. Sah, Homology of Euclidean groups of motions made discrete and Euclidean scissors congruences, *Acta Math.* **164** (1990), 1–27.

C. F. Gauss, *Werke, Bd. 8*, p. 241, p. 244.

A. B. Goncharov, Volumes of hyperbolic manifolds and mixed Tate motives, *J. Amer. Math. Soc.* **12** (1999), 569–618.

B. H. Gross, *On the Values of Artin L-functions*, (c. 1980), unpublished.

U. Haagerup and H. Munkholm, Simplices of maximal volume in hyperbolic *n*-space, *Acta Math.* **147** (1981), 1–11.

D. Hilbert, *Gesammelte Abhandlungen, Bd. 3*, Chelsea, 1965, 301–302.

H. Hopf, *Differential Geometry in the Large*, LNM 1000, Springer, 1983.

W. H. Jackson, Wallace's theorem concerning plane polygons of the same area, *Amer. J. Math.* **34** (1912), 383–390.

B. Jessen, A remark on the volume of polyhedra, *Mat. Tidsskr. B* (1941), 59–65. (In Danish.)

B. Jessen, The algebra of polyhedra and the Dehn-Sydler theorem, *Math. Scand.* **22** (1968), 241–256.

B. Jessen, Zur Algebra der Polytope, *Göttingen Nachr. Math. Phys.* (1972), 47–53.

B. Jessen, J. Karpf and A. Thorup, Some functional equations in groups and rings, *Math. Scand.* **22** (1968), 257–265.

L. Lewin, *Dilogarithm and Associated Functions*, Macdonald, 1958.

L. Lewin, *Polylogarithms and Associated Functions*, Elsevier, 1981.

L. Lewin, ed., *Structural Properties of Polylogarithms*, Math. Surveys and Monographs, 37, Amer. Math. Soc., Providence, 1991.

J. Mather, *A letter to C. H. Sah*, 1975.

J. W. Milnor, *How to Compute Volume in Hyperbolic Space*, Collected Papers, I, Publish or Perish, 1994.

J. W. Milnor, *Hyperbolic Geometry: The First 150 Years*, Collected Papers, I, Publish or Perish, 1994.

W. D. Neumann and J. Yang, Problems for K-theory and Chern-Simons invariants of hyperbolic 3-manifolds, *L'Ens. Math.* **41** (1995), 281–296.

W. D. Neumann and J. Yang, Invariants from triangulation for hyperbolic 3-manifolds, *Elec. Res. Announce. of Amer. Math. Soc.* **1** (1995), 72–79.

W. D. Neumann and J. Yang, Bloch invariants of hyperbolic 3-manifolds, *Duke Math. J.* **96** (1999), 29–59.

W. Parry Note on universal odd Kubert function, (c. 1984), unpublished.

C. H. Sah, *Hilbert's Third Problem: Scissors Congruence,* Res. Notes in Math. 33, Pitman, 1979.

C. H. Sah, Scissors congruence, I, Gauss-Bonnet map, *Math. Scand.* **49** (1981), 181–210.

C. H. Sah, Homology of Lie groups made discrete, I, *Comment. Math. Helv.* **61** (1986), 308–347.

C. H. Sah, Homology of Lie groups made discrete, III, *J. Pure and Appl. Algebra* **56** (1989), 269–312.

C. H. Sah and J. B. Wagoner, Second homology of Lie groups made discrete, *Comm. in Algebra* **5** (1977), 611–642.

L. Schläfli, On the multiple integral..., *Quart J. Math.* **3** (1860), 54–68, 97–108, see also *Gesam. Math. Abhand.* **1**, Birkhäuser, 1950.

A. A. Suslin, *Homology of GL_n, Characteristic Classes and Milnor K-theory,* Springer LNM, 1046 (1984), 357–375.

A. A. Suslin, Algebraic K-theory of fields, *Proc. ICM,* I, Berkeley, (1986), 222–244.

A. A. Suslin, K_3 of a field and the Bloch group, *Proc. Steklov Inst. of Math.* (1991), 217–239.

J. P. Sydler, Conditions nécessaires et suffisantes pour l'équivalence des polyèdres l'espace euclidien à trois dimensions, *Comment. Math. Helv.* **40** (1965), 43–80.

A. Thorup, Om delinger af polyedre, (c. 1970), unpublished notes in Danish.

W. P. Thurston, *The Geometry and Topology of Three-Manifolds,* Geometry Center, Univ. of Minn., 1979.

W. Wallace, Question 269, in *Thomas Leyborne, Math. Repository,* III, London, 1814.

V. B. Zylev, Equicomposability of equicomplementable polyhedra, *Sov. Math. Doklady* **161** (1965), 453–455.

V. B. Zylev, *G*-composedness and *G*-complementability, *Sov. Math. Dok-lady* **179** (1968), 403–404.

Johan L. Dupont   Chih-Han Sah[†] (deceased)
Matematisk Institut   Department of Mathematics
Aarhus University   SUNY at Stony Brook
Ny Munkegade, DK-8000   Stony Brook, NY 11794-3651
Århus C, Denmark
e-mail: dupont@mi.aau.dk

AMS Subject Classifications: Primary 53B45; Secondary 20G10, 22E99, 51M10, 52B10

Poisson Structure for Restricted Lie Algebras

V. Kac and A. Radul

Introduction

The Poisson bracket is an important invariant of a deformation of a commutative associative algebra A (cf. [D]). Given a family of associative (but not necessarily commutative) algebras A_h, such that $A_0 = A$, one defines a Poisson bracket of two elements $a, b \in A$ by taking $a(h)$, $b(h) \in A_h$, such that $a(0) = a$, $b(0) = b$ and letting

$$\{a, b\} = \lim_{h \to 0} \frac{[a(h), b(h)]}{h}. \tag{1}$$

If A is not commutative, then (1) defines a Poisson structure on the center of A and, more generally, central elements define Poisson derivatives of A. This has been used extensively in the recent work on the representation theory of quantum groups at roots of 1 [DKP].

On the other hand, the representation theory of quantum groups at roots of 1 is parallel to that of restricted Lie algebras in characteristic p [WK]. This is not surprising in view of the fact that reduction mod p of a quantum group at the p-th root of 1 produces essentially the universal enveloping algebra of a restricted Lie algebra of classical type (cf. [L]).

It is therefore natural to ask what the Poisson structure is on the center of the enveloping algebra of a restricted Lie algebra, and what are the corresponding Poisson derivations.

In §1 we construct the Poisson derivations in a general algebraic setup. In §2 we recall the necessary material on restricted Lie algebras, and in §3 we calculate the Poisson bracket on the center of the enveloping algebra of a restricted Lie algebra and the Poisson derivations.

We explore here neither connections to quantum groups nor applications of the Poisson structure to the representation theory of restricted Lie algebras, leaving this for future publication.

1   General setup

Let R be a commutative associative ring, let $\mathfrak{m}$ be a principal ideal of R, and let $k = R/\mathfrak{m}$. Let U be a unital associative algebra over R; assume that U is a torsion free $\mathfrak{m}$-module. Denote by $U_\mathfrak{m}$ the algebra $U/\mathfrak{m}U$ over k and let $\varphi : U \to U_\mathfrak{m}$ be the canonical homomorphism. Let $Z_\mathfrak{m}$ be the center of $U_\mathfrak{m}$ and let $N^\mathfrak{m} = \varphi^{-1}(Z_\mathfrak{m}) \subset U$.

Fix a generator ϵ of the ideal $\mathfrak{m}$. Given $a \in N^\mathfrak{m}$ and $b \in U$, we have $[a, b] = \epsilon b_1$, where $b_1 \in U$ is uniquely defined; we write $b_1 = \frac{[a,b]}{\epsilon}$. Hence we may define a *Poisson derivation* P_a of $U_\mathfrak{m}$ by the formula

$$P_a(u) = \varphi\left(\frac{[a, \varphi^{-1}(u)]}{\epsilon}\right), \quad u \in U_\mathfrak{m}. \tag{2}$$

This is clearly independent of the choice of the preimage $\varphi^{-1}(u) \in U$ of u, and it is easy to check that P_a is indeed a derivation of the k-algebra $U_\mathfrak{m}$. Equally straightforward are the following properties ($u \in U_\mathfrak{m}$):

$$P_{ab}(u) = P_a(u)\varphi(b) + \varphi(a)P_b(u), \quad a, b \in N^\mathfrak{m}; \tag{3}$$
$$P_{\epsilon^{-1}[a,b]} = [P_a, P_b], \quad a, b \in N^\mathfrak{m}; \tag{4}$$
$$P_{\epsilon a}(u) = [\varphi(a), u], \quad a \in U \tag{5}$$

Recall the following well-known

Lemma 1 *Let D be a derivation of an algebra U. Then the center of U is D-invariant.*

Proof.   If $a \in U$ and z is a central element, we have $az = za$. Applying D to both sides we have $D(a)z + aD(z) = D(z)a + zD(a)$. Since z is central we deduce that $aD(z) = D(z)a$. □

Thus, we may define a Poisson bracket $\{\ ,\ \}$ on the center $Z_\mathfrak{m}$ of $U_\mathfrak{m}$ by the formula

$$\{a, b\} = P_{\varphi^{-1}(a)}(b) = -P_{\varphi^{-1}(b)}(a) = \varphi\left(\frac{[\varphi^{-1}(a), \varphi^{-1}(b)]}{\epsilon}\right). \tag{6}$$

It is clear that this bracket is skew-symmetric, it follows from (4) that this bracket satisfies the Jacobi identify and (3) implies the Leibniz rule

$$\{ab, c\} = a\{b, c\} + \{a, c\}b.$$

Remark 1   Define a Lie algebra structure on $N_m := N^m/m^2U$ by $\{a, b\} = \epsilon^{-1}[a, b]$. The map $\alpha : U \to N^m$, given by $u \mapsto \epsilon u$, induces a Lie algebra injective homomorphism $\overline{\alpha} : U_m \to N_m$. The map $\varphi : N^m \to Z_m$ induces a Lie algebra surjective homomorphism $\overline{\varphi} : N_m \to Z_m$. We have an exact sequence of Lie algebras

$$0 \longrightarrow U_m \overset{\overline{\alpha}}{\longrightarrow} N_m \overset{\overline{\varphi}}{\longrightarrow} Z_m \longrightarrow 0. \tag{7}$$

Thus we may say that every element of Z_m defines a Poisson derivation of the k-algebra U_m defined up to addition of inner derivations of U_m.

2   Restricted Lie algebras

Let k be a field of characteristic $p > 0$. Let $\mathfrak{g}$ be a Lie algebra over k, and let $U(\mathfrak{g})$ denote its universal enveloping algebra. Recall that a *p-structure* on $\mathfrak{g}$ is a map $a \mapsto a^{[p]}$ of $\mathfrak{g}$ into itself satisfying the following three axioms [J]:

$$(\alpha a)^{[p]} = \alpha^p a^{[p]}, \quad \alpha \in k, \tag{8}$$
$$(\operatorname{ad} a)^p = \operatorname{ad} a^{[p]}, \tag{9}$$
$$(a + b)^{[p]} = a^{[p]} + b^{[p]} + \wedge(a, b), \tag{10}$$

where $\wedge(a, b) \in \mathfrak{g}$ is defined by

$$\wedge(a, b) = \sum_{j=1}^{p-1} \wedge_j(a, b), \quad \sum_{j=1}^{p-1} j \wedge_j (a, b) \lambda^{j-1} = \operatorname{ad}(\lambda a + b)^{p-1} a. \tag{11}$$

A Lie algebra with a p-structure is called a *restricted* Lie algebra.

Note that for the existence of a p-structure on $\mathfrak{g}$, it is necessary that all derivations $(\operatorname{ad} a)^p$ of $\mathfrak{g}$ ($a \in \mathfrak{g}$) be inner. In fact, this is sufficient; for choosing a basis $\{a_i\}$ of $\mathfrak{g}$, we have $(\operatorname{ad} a_i)^p = \operatorname{ad} a_i^{[p]}$ for some $a_i^{[p]} \in \mathfrak{g}$, and this extends uniquely to a p-structure on $\mathfrak{g}$ by (8) and (10). Note also that two p-structures on $\mathfrak{g}$ differ by a semi-linear map $\varphi : \mathfrak{g} \to \operatorname{center}(\mathfrak{g})$. Here and further by semi-linear map φ we mean

$$\varphi(a + b) = \varphi(a) + \varphi(b) \quad \text{and} \quad \varphi(\alpha a) = \alpha^p \varphi(a).$$

Example 1 [J]   Let A be an associative algebra over k viewed as a Lie algebra with the usual bracket. Then $a^{[p]} = a^p$ is a p-structure.

Example 2 [B] Let G be an affine algebraic group over k and let $\mathfrak{g}$ be its Lie algebra, i.e., Lie algebra of all derivations D of the coordinate ring of G which commute with left translations in G. Then $D^{[p]} = D^p$ is a p-structure.

The following lemma is well-known.

Lemma 2 *Let $\mathfrak{g}$ be a restricted Lie algebra over k. Then*

(a) *For $a \in \mathfrak{g}$ the element $a^p - a^{[p]}$ lies in the center of $U(\mathfrak{g})$.*

(b) *[WK] The map $\mathfrak{g} \to$ center $U(\mathfrak{g})$ defined by $a \mapsto a^p - a^{[p]}$ is semilinear.*

Proof. (a) follows from
$$(\operatorname{ad} a)^p = \operatorname{ad} a^p \tag{12}$$
(which holds by the Leibniz formula) and (9).
    (b) follows from (8), (10) and Example 1.   □

Lemma 3 *Let $\mathfrak{g}$ be a restricted Lie algebra over k. Let $Y = \{y_1, \ldots, y_p\}$ be a collection of p elements. Then*
$$\sum_{\sigma \in S_p} y_{\sigma(1)} \cdots y_{\sigma(p)} \in \mathfrak{g}. \tag{13}$$

Proof. Due to Example 1 we have
$$(\lambda_1 y_1 + \ldots + \lambda_p y_p)^p = \lambda_1^p y_1^p + \ldots + \lambda_p^p y_p^p + S(\lambda), \tag{14}$$
where $S(\lambda) \in \mathfrak{g}$. But the expression in (13) is the coefficient of $\lambda_1 \ldots \lambda_p$ in (14). Hence it lies in $\mathfrak{g}$.   □

Remark 2 Given an element x of a Lie algebra $\mathfrak{g}$ with a p-structure, we may define inductively for $a \in \mathfrak{g}$,
$$a^{[p^k]} = (a^{[p^{k-1}]})^{[p]} \quad \text{for } k \geq 1, \quad a^{[0]} = a.$$

Then $a^{p^k} - a^{[p^k]}$ is a central element, which can be expressed via the elements $x^p - x^{[p]} (x \in \mathfrak{g})$ using the formula
$$a^{p^k} - a^{[p^k]} = (a^p - a^{[p]})^{p^{k-1}} + \left((a^{[p]})^{p^{k-1}} - (a^{[p]})^{[p^{k-1}]} \right). \tag{15}$$

The latter formula is immediate by the remark that for a central element $z = a - b$, one has $z^{p^k} = a^{p^k} - b^{p^k}$ (indeed: $a^{p^k} = (z+b)^{p^k} = z^{p^k} + b^{p^k}$).

3 Poisson structure for restricted Lie algebras

Let $\mathfrak{g}$ be a Lie algebra over $\mathbb{Z}$ and let U be its universal enveloping algebra (over $\mathbb{Z}$). For a prime $p \in \mathbb{Z}$, let $\mathfrak{g}_p = \mathfrak{g}/p\mathfrak{g}$ be the Lie algebra over the field $\mathbb{F}_p = \mathbb{Z}/p\mathbb{Z}$, and let $U_p = U/pU$ be the universal enveloping algebra of $\mathfrak{g}_p$. Denote by Z_p the center of U_p. Given $u \in U$, we denote by u_p its image in U_p.

Let $a \in \mathfrak{g} \subset U$ be such that there exists $a' \in \mathfrak{g}$ for which

$$(\operatorname{ad} a)^p x \equiv [a', x] \bmod p \quad \text{for all } x \in \mathfrak{g}. \tag{16}$$

Then since (see §2)

$$(\operatorname{ad} a)^p x \equiv [a^p, x] \bmod p \quad \text{for all } x \in \mathfrak{g}, \tag{17}$$

we have

$$a_p^p - a_p' \in Z_p. \tag{18}$$

Hence as in §1, we may consider the Poisson derivation $D_a = P_{a^p - a'}$ of U_p:

$$D_a(u_p) = \left(\frac{[a^p - a', u]}{p!} \right)_p.$$

Here and further u_p denotes an element of U_p and u denotes a preimage of u_p in U. Due to (17) we may also introduce the following k-endomorphism of U_p:

$$B_a(u_p) = \left(\frac{((\operatorname{ad} a)^p - \operatorname{ad} a')u}{p!} \right)_p.$$

Note that changing the choice of a' in (16) adds to D_a and B_a (the same) inner derivation.

Lemma 4 *One has*

$$D_a(u_p) = B_a(u_p) + \sum_{s=1}^{p-1} \frac{(-1)^s}{s} \left((\operatorname{ad} a_p)^s u_p \right) a_p^{p-s}, \quad u_p \in U_p.$$

Proof. Recall that in any associative algebra, one has

$$a^p u = \sum_{s=0}^{p} \binom{p}{s} \left((\operatorname{ad} a)^s u \right) a^{p-s}.$$

Hence

$$[a^p, u] = (\operatorname{ad} a)^p u + p! \sum_{s=1}^{p-1} \frac{(-1)^s}{s} \left((\operatorname{ad} a)^s u \right) a^{p-s}. \qquad \square$$

Lemma 5 *Suppose that* $\mathfrak{g}_p$ *is a restricted Lie algebra. Then one has for some* $g \in \mathfrak{g}$,

$$(\operatorname{ad} a)^p u^p \equiv \sum_{s=0}^{p-1} u^s \left((\operatorname{ad} a)^p u \right) u^{p-s-1} + p! [a, u]^p + g \bmod p^2.$$

Proof. Let $u(j_1, \ldots, j_s) = ((\operatorname{ad} a)^{j_1} u) \cdots ((\operatorname{ad} a)^{j_s} u)$. By the Leibniz formula, we have

$$(\operatorname{ad} a)^p u^p = \sum_{\substack{0 \le j_k < p \\ j_1 + \ldots j_p = p}} \frac{p!}{j_1! \ldots j_p!} u(j_1, \ldots, j_p) + \sum_{s=0}^{p-1} u^s u(p) u^{p-s-1}. \tag{19}$$

Denote by $M(s_0, s_1, \ldots)$ the set of p-tuples of integers consisting of s_0 zeros, s_1 1's, etc., where $\sum_{j=0}^{p} j s_j = p$, and pick $(\lambda_1, \ldots, \lambda_p) \in M(s_0, s_1, \ldots)$. Then we have

$$\sum_{(j_1, \ldots, j_p) \in M(s_0, s_1, \ldots)} u(j_1, \ldots, j_p) \prod_j s_j! = \sum_{\sigma \in S_p} u(\lambda_{\sigma(1)}, \ldots, \lambda_{\sigma(p)}).$$

From this and Lemma 5 it follows that $\bmod p^2$, the only nontrivial contribution from the first sum of (19), comes from $u(1, \ldots, 1) = [a, u]^p$. □

We can prove now the key result:

Theorem 1 *Let* $\mathfrak{g}$ *be a Lie algebra over* $\mathbb{Z}$, *such that* $\mathfrak{g}_p$ *is a restricted Lie algebra over* $\mathbb{F}_p$. *Then for any* $a_p, b_p \in \mathfrak{g}_p$, *there exists* $g(a, b) \in \operatorname{center}(\mathfrak{g}_p)$ *such that*

$$\left\{ a_p^p - a^{[p]}, \, b_p^p - b^{[p]} \right\} := D_a \left(b_p^p - b_p^{[p]} \right) = [a, b]_p^p - [a, b]_p^p + g(a, b). \tag{20}$$

Proof. Since the element $b_p^p - b_p^{[p]}$ is central, by Lemma 4 we need to prove that

$$\left((\operatorname{ad} a)^p - \operatorname{ad} a^{[p]} \right) (b^p - b^{[p]}) - p! [a, b]^p \in \mathfrak{g} \bmod p^2. \tag{21}$$

By Lemma 5, (21) will follow from

$$\sum_{s=0}^{p-1} b^s \left(((\operatorname{ad} a)^p - \operatorname{ad} a^{[p]}) b \right) b^{p-s-1} \in \mathfrak{g} \bmod p^2. \tag{22}$$

On the other hand, $((\operatorname{ad} a_p)^p - \operatorname{ad} a_p^{[p]})b_p = 0$, hence $((\operatorname{ad} a)^p - (\operatorname{ad} a^{[p]}))b = p!\, b_1$ for some $b_1 \in \mathfrak{g}$. Hence (22) will follow from

$$\sum_{s=0}^{p-1} b^s b_1 b^{p-s-1} \in \mathfrak{g} \bmod p.$$

But this follows from Lemma 3. Finally, $g(a, b)$ is a central element of $\mathfrak{g}_p$ due to Lemmas 1 and 2a. $\qquad\square$

Remark 3 Using similar calculations, it is easy to show that center$(\mathfrak{g}_p)$ is invariant with respect to the derivations D_a and that this induces a representation of $\mathfrak{g}_p$ in center$(\mathfrak{g}_p)$. The bi-semilinear map $g : \mathfrak{g}_p \times \mathfrak{g}_p \to$ center$(\mathfrak{g}_p)$ is a 2-cocycle for this representation.

Let G be an affine algebraic group defined over $\mathbb{Z}$. Then its Lie algebra Lie G is a Lie algebra over $\mathbb{Z}$ with the property that the Lie algebra (Lie $G)_p$ has a canonical p-structure for any prime p (cf. Example 2).

Corollary 1 *For the restricted Lie algebra* (Lie $G)_p$ *one has*

$$\left\{ a_p^p - a_p^{[p]},\ b_p^p - b_p^{[p]} \right\} = [a, b]_p^p - [a, b]_p^{[p]}.$$

Proof. We may assume that $G \subset SL_n$ is an algebraic subgroup, so that the p-structure on $\mathfrak{g}_p$ is induced from that on $sl_n(\mathbb{F}_p)$. We also may assume that $(n, p) = 1$. But then center$(sl_n(\mathbb{F}_p)) = 0$, hence the 2-cocyle $g(a, b) = 0$, and the corollary follows from Theorem 1. $\qquad\square$

Remark 4 Let $\mathfrak{g}$ be a Lie algebra over $\mathbb{Z}$ such that $\mathfrak{g}_p$ is a restricted Lie algebra. Denote by D_p (resp. Z_0) the linear span in U_p of all the elements a and a^p (resp. $a^p - a^{[p]}$), where $a \in \mathfrak{g}_p$. Then the exact sequence (7) induces the exact sequence of Lie algebras

$$0 \longrightarrow \mathfrak{g}_p \longrightarrow D_p \longrightarrow Z_0 \longrightarrow 0.$$

Remark 5 One can calculate the Poisson bracket of elements of the form $a^{p^k} - a^{[p^k]}(a \in \mathfrak{g}_p)$ using Remark 2 and (20):

$$\begin{aligned}
\{a^{p^k} - a^{[p^k]},\ b^{p^s} - b^{[p^s]}\}& \\
= [a^{[p^{k-1}]}, b^{[p^{s-1}]}]^p &- [a^{[p^{k-1}]}, b^{[p^{s-1}]}]^{[p]} + g(a^{[p^{k-1}]}, b^{[p^{s-1}]}).
\end{aligned}$$

References

[B] A. Borel, *Linear Algebraic Groups*, Benjamin, New York, 1969.

[DKP] C. De Concini, V. G. Kac, and C. Procesi, Quantum coadjoint action, *Amer. Math. Soc.* **5** (1992), 151–189.

[D] V. G. Drinfeld, Quantum groups, *Proc. ICM*, Berkeley **1** (1986), 789–820.

[J] N. Jacobson, *Lie Algebras*, Interscience Publ., 1962.

[L] G. Lusztig, Finite-dimensional Hopf algebras arising from quantum groups, *J. Amer. Math. Soc.* **3** (1990), 257–296.

[WK] B. Yu. Weisfeiler, V. G. Kac, On irreducible representations of Lie p-algebras, *Funct. Anal. Appl.* **5:2** (1971), 28–36.

Victor Kac
Department of Mathematics
M.I.T.
Cambridge, MA 02139
e-mail:kac@math.mit.edu

AMS Subject Classification: 17B67

Noncommutative Smooth Spaces

Maxim Kontsevich and Alexander L. Rosenberg

Conventions and notations

We will work in the category Alg_k of associative unital algebras over a fixed base field k. If $A \in Ob(Alg_k)$, we denote by $1_A \in A$ the unit in A and by $m_A : A \otimes A \longrightarrow A$ the product. For an algebra A, we denote by A^{opp} the opposite algebra, i.e., the same vector space as A endowed with the multiplication $m_{A^{opp}}(a \otimes b) := m_A(b \otimes a)$. If A and B are two algebras, then $A \otimes_k B$ is again an algebra. Also, $A \star B$ denotes the free product of A and B over k, the coproduct in the category Alg_k. By A-mod we denote the abelian category of left A-modules. Analogously, mod-A are right modules (the same as A^{opp}-modules) and A-mod-A are bimodules over k, or, equivalently, $A \otimes_k A^{opp}$-modules. We shall write $\otimes$ instead of $\otimes_k$. For a vector space V, we denote by $Sym^*(V)$ and $\otimes^*(V)$ resp. the free commutative associative (polynomial) and free associative (tensor) k-algebra respectively, generated by V.

1 Smooth algebras

A typical example of an algebra for this paper is a free finitely generated algebra $k\langle x_1, \ldots, x_d \rangle$, in contrast to the usual noncommutative algebras that are close to commutative ones, such as the universal enveloping algebras, algebras of polynomial differential operators, and so on. We consider the free algebra as "the algebra of functions on the non-commutative affine space," which we denote by $N\mathbb{A}^d$. Recently there have been several attempts to understand the algebraic geometry of this space. Gelfand and Retakh started the study of basic identities in the free skew field with d generators (see [GR]), which can be considered as the "generic point" in $N\mathbb{A}^d$. One of us (see [K]) tried to develop "differential geometry" in the algebra of the noncommutative formal power series, i.e., at the formal neighborhood of $N\mathbb{A}^d$ at zero. Kapranov (see [Ka]) described a differential-geometric picture related to the completion of the free algebra with respect to the commutator filtration, i.e., to the algebra of functions on the infinitesimal neighborhood of the usual affine space $\mathbb{A}^d$ in $N\mathbb{A}^d$. Here we would like to study $N\mathbb{A}^d$ as the whole space, without completions and localizations. The basic intrinsic property of the free algebra is *smoothness*.

1.1 Equivalent definitions

A few years ago J. Cuntz and D. Quillen gave the following

Definition ([CQ1], Definition 3.3 and Proposition 6.1). An algebra A is quasi-free (= formally smooth) iff it satisfies one of the following equivalent properties:

1) (Lifting property for nilpotent extensions) for any algebra B, a two-sided nilpotent ideal $I \subset B$ ($I = BIB$, $I^n = 0$ for $n \gg 0$), and for any algebra homomorphism $f : A \longrightarrow B/I$, there exists an algebra homomorphism $\tilde{f} : A \longrightarrow B$ such that $f = pr_{B \longrightarrow B/I} \circ \tilde{f}$, where $pr_{B \longrightarrow B/I} : B \longrightarrow B/I$ is the natural projection.

2) $Ext^2_{A-mod-A}(A, M) = 0$ for any bimodule $M \in Ob(A\text{-}mod\text{-}A)$.

3) The A-bimodule $\Omega^1_A := Ker(m_A : A \otimes A \longrightarrow A)$ (see 1.1.2) is projective.

The definition of formal smoothness via the lifting property 1) is analogous to Grothendieck's (actually Quillen's) definition of formally smooth algebras in the commutative case. The equivalence of the properties (1)–(3) is an easy exercise in homological algebra.

Properties (1) and (2) are not constructive: one cannot check them directly. The property (3) looks better, although it is still not clear *a priori* how to work with it in complicated infinite-dimensional examples. J. Cuntz and D. Quillen found another characterization of smooth algebras which is convenient for calculations in practice. In order to give this characterization, we need an auxiliary definition.

Let TA be the algebra generated by symbols $\underline{a}, \underline{Da}$, where $a \in A$, subject to the following relations:

a) the map $a \mapsto \underline{a}$ is a homomorphism of unital k-algebras,

b) the map $a \mapsto \underline{Da}$ is k-linear,

c) $\underline{D(a \cdot b)} = \underline{a} \cdot \underline{Db} + \underline{Da} \cdot \underline{b}$, (the Leibniz rule).

The algebra TA is naturally $\mathbb{Z}_{\geq 0}$-graded: $deg(\underline{a}) = 0$, $deg(\underline{Da}) = +1$. We identify A with the subalgebra of TA consisting of elements of degree zero.

1.1.1 Differential envelope

Before going further, we would like to note that TA is isomorphic as an abstract $\mathbb{Z}$-graded algebra to the so-called *differential envelope* ΩA, the universal differential $\mathbb{Z}$-graded super-algebra containing A.

By definition, ΩA is the universal $\mathbb{Z}$-graded super-algebra containing A and endowed with an odd differential d of degree $+1$ (satisfying the Leibniz rule in the super sense of $d(a \cdot b) = d(a) \cdot b + (-1)^{deg(a)} a \cdot d(b)$) and such that $d^2 = 0$. One can see immediately that $d1 = 0$. As a vector space, the n-th graded component $(\Omega A)^n$ for $n \geq 0$ is isomorphic to $A \otimes (A/k \cdot 1_A)^{\otimes n}$. The isomorphism is given by the following map

$$a_0 \otimes a_1 \otimes \ldots \otimes a_n \longmapsto a_0 \cdot da_1 \cdot \ldots \cdot da_n \in (\Omega A)^n.$$

Furthermore, the space of 1-differentials $\Omega_A^1 := (\Omega A)^1$ can be identified via the map

$$A^{\otimes 3} \longrightarrow \Omega_A^1, \quad a_1 \otimes a_2 \otimes a_3 \longmapsto a_a \cdot da_2 \cdot a_3$$

with the quotient space $A^{\otimes 3}/\partial_4(A^{\otimes 4})$, where

$$\partial_4(a_1 \otimes a_2 \otimes a_3 \otimes a_4) = a_1 a_2 \otimes a_3 \otimes a_4 - a_1 \otimes a_2 a_3 \otimes a_4 + a_1 \otimes a_2 \otimes a_3 a_4).$$

The map ∂_4 can be, evidently, included into one of standard complexes in homological algebra

$$\ldots \longrightarrow A^{\otimes 4} \longrightarrow A^{\otimes 3} \longrightarrow A^{\otimes 2} \longrightarrow A \longrightarrow 0$$

This complex can be contracted using the unit element $1_A \in A$, and thus has vanishing cohomology. Therefore $Cokernel(\partial_4) = Kernel(\partial_2)$, and we get an equivalent description of Ω_A^1 as the kernel of the multiplication map $A \otimes A \longrightarrow A$.

If one considers analogous definitions in the purely commutative case, one can see immediately the difference between TA and ΩA. In the case of A equal to the algebra of functions on a smooth affine variety X, the commutative analogue of the algebra TA is the algebra of functions on the total space TX of the tangent bundle to X. On the contrary, the commutative analogue of ΩA is the algebra of differential forms on X, which is the same as the algebra of functions on the supermanifold ΠTX, the total space of the odd tangent bundle to Z (as usual in super-mathematics, the letter Π denotes the functor for changing the parity).

1.1.2   Another criterion of formal smoothness

Theorem   *An algebra A is formally smooth iff it satisfies the property that*

4) there exists a derivation $\underline{D} : TA \longrightarrow TA$ of degree $+1$ such that $\underline{D}(\underline{a}) = \underline{Da}$ for all $a \in A$.

If A is an algebra generated by elements $(a_i)_{i \in I}$, then in order to define $\underline{D}$ one should define only elements $\underline{D}(\underline{Da_i}) \in TA$ of degree $+2$ satisfying certain relations. In the case when the generating set I is finite and all relations between (a_i) follow from a finite number of relations, we say that A is finitely presented. It is easy to see that in this case algebra TA is also finitely presented, and in order to check that some candidates for $\underline{D}(\underline{Da_i})$ satisfy needed relations, one should make only a finite calculation.

The theorem analogous to the one above holds in the commutative case (if the characteristic of the ground field k is zero). The derivation $\underline{D}$ in the commutative case is a vector field on the total space of the tangent bundle. One can easily see that choices of $\underline{D}$ correspond to symmetric connections (i.e., connections with vanishing torsion) on the tangent bundle to $Spec(A)$.

1.1.3   Homological properties of modules over smooth algebras

Theorem ([CQ1], Proposition 5.1)   *If A is a formally smooth algebra, then A is hereditary, i.e., the category A-mod has homological dimension ≤ 1. In other words, any submodule of a projective module is projective. Equivalently, every A-module admits a projective resolution $P_{-1} \longrightarrow P_0$ of length 2.*

The proof is immediate because

$$Ext^n_{A-mod}(\mathcal{E}, \mathcal{F}) = Ext^n_{A-mod-A}(A, Hom_{k-mod}(\mathcal{E}, \mathcal{F}))$$
$$\forall \mathcal{E}, \mathcal{F} \in Ob(A - mod).$$

Definition   We call an algebra A smooth, if it is formally smooth and finitely generated.

Theorem   *If A is a smooth algebra and $\mathcal{E}, \mathcal{F}$ are A-modules finite-dimensional over k, then the vector spaces $Ext^0_{A-mod}(\mathcal{E}, \mathcal{F}) = Hom_{A-mod}(\mathcal{E}, \mathcal{F})$ and $Ext^1_{A-mod}(\mathcal{E}, \mathcal{F})$ are both finite-dimensional.*

The statement of the theorem is evident for Ext^0. The space Ext^1 coincides with the set of equivalence classes of structures of an A-module on $\mathcal{E} \oplus \mathcal{F}$ such that natural maps

$$\mathcal{F} \longrightarrow \mathcal{E} \oplus \mathcal{F} \longrightarrow \mathcal{E}$$

are morphisms of A-modules, modulo the action of the vector space $Hom_{k-mod}(\mathcal{E}, \mathcal{F})$ considered as an abelian group. If A is finitely generated, then the set of A-module structures on $\mathcal{E} \oplus \mathcal{F}$ as above is a finite-dimensional vector space over k, and we get the statement of the theorem.

1.2 Examples of smooth algebras

E1) The free algebra $k\langle x_1, \ldots, x_d \rangle = \otimes^*(k^d)$ $d \geq 0$. We think of this algebra as the one corresponding to a noncommutative affine space $N\mathbb{A}_k^d$ (see Section 3 where we define noncommutative spaces in general).

E2) The matrix algebra $Mat(n \times n, k)$.

E3) The algebra of upper-triangular matrices

$$A = \{(a_{ij})_{1 \leq i,j \leq n} | \ a_{ij} \in k, \ a_{ij} = 0 \ \text{ for } \ i > j\} \subset Mat(n \times n, k),$$

E4) $\mathcal{O}(C)$, the algebra of functions on a smooth affine curve C over k.

E5) $Paths(\Gamma)$, the algebra of finite paths in a finite oriented graph Γ. The basis of $Paths(\Gamma)$ consists of sequences $(v_0, e_{0,1}, v_1, e_{1,2}, \ldots, e_{n-1,n}, v_n)$, $n \geq 0$ of vertices v_i of Γ and oriented edges $e_{i,i+1}$ connecting v_i with v_{i+1} (there could be several edges connecting two given vertices). The unit element in $Paths(\Gamma)$ is equal to the sum over vertices v of Γ of paths of zero length (v). The multiplication in $Paths(\Gamma)$ is given by the concatenation of paths:

$$(v_0, e_{0,1}, \ldots, v_n) \cdot (v'_0, e'_{0,1}, \ldots, v'_{n'})$$
$$= (v_0, e_{0,1}, \ldots, v_n, e'_{0,1}, \ldots, v'_{n'})$$

if $v_n = v'_0$, and 0 otherwise.

The last example E5) contains E1) and E3) as particular cases. In E1) the corresponding graph has one vertex and d loops, and in E3) it is the chain of $n-1$ consecutive oriented edges.

There are several constructions, from which one can use whatever one can construct new smooth algebras from the old ones. In the following list, A, A_1, A_2 denote smooth algebras, and the result of the constructions is automatically a smooth algebra:

C1) $A_1 \oplus A_2$, the direct sum of two smooth algebras (or an arbitrary finite number of smooth algebras),

C2) $A_1 \star A_2$, the free product of smooth algebras.

C3) (Localization) If S is a subset of a formally smooth algebra A, then the algebra $A[S^{-1}]$, obtained by formally adjoining the inverses of elements of S, is formally smooth. In particular, if A is smooth and $S \subset A$ is an arbitrary finitely generated multiplicative subset of A, then $S^{-1}A$ is smooth. More generally, one can invert not only individual elements of A but square matrices with coefficients in A (i.e., S could be a subset of $\coprod_{n \geq 1} Mat(n \times n, A)$), or even rectangular matrices.

C4) $End_{A-mod}(P)$, where P is a finitely-generated projective A-module.

C5) (The total space of the tangent bundle) TA, the algebra defined in 1.1.

1.3   Representation spaces of smooth algebras

Let A be a smooth k-algebra. We associate with A an infinite sequence $(Repr_n^A, n \geq 1)$ of smooth affine varieties over k. The n-th variety $Repr_n^A$ parameterizes homomorphisms from A to the standard matrix algebra $Mat(n \times n, k)$. The set of B-points of it, where B is a *commutative* algebra over k, is defined as

$$Repr_n^A(B) := Hom_{Alg_k}(A, Mat(n \times n, B)).$$

It is clear that since A is finitely-generated, the scheme $Repr_n^A$ is an affine scheme of finite type over k. Moreover, from the definitions of smoothness in both the commutative and noncommutative case, it follows that the scheme $Repr_n^A$ is smooth. The affine algebraic group $PGL_k(n)$ acts via conjugations on $Repr_n^A$.

Every element $a \in A$ gives tautologically a matrix-valued function $\hat{a} \in \mathcal{O}(Repr_n^A) \otimes Mat(n \times n, k)$ on $Repr_n^A$.

1.3.1   Examples

For $A = k\langle x_1, \ldots, x_d \rangle$, the representation scheme $Repr_n^A$ coincides with the affine space $\mathbb{A}^{dn^2}$ over k.

For $A = k \oplus k = k\langle p \rangle/(p^2 = p)$, the scheme $Repr_n^A$ is the disjoint union over all integers m, $0 \leq m \leq n$, of certain bundles over Grassmannians $Gr(m, n)$. Fibers of these bundles are affine spaces parallel to the fibers of the cotangent bundles. The component of $Repr_n^A$ corresponding to m (— the rank of the image of $p \in A$) has dimension $2m(n - m)$.

The example of $A = k \oplus k$ has an interesting feature. One can see that $Repr_n^A$ carries a natural non-degenerate closed 2-form

$$\omega = Trace(\hat{p}\, d\hat{p} \wedge d\hat{p})$$

where $\hat{p}$ is the matrix-valued function on $Repr_n^A$ corresponding to the generator p of A.

1.3.2 Noncommutative analogues
of differential-geometric structures

We would like to propose the following general principle which could help to look for noncommutative versions of usual geometric notions (see also [K]).

A noncommutative structure of some kind on A should give an analogous "commutative" structure on all schemes $Repr_n^A$, $n \geq 1$.

Here we make a list of several natural candidates:

Functions Elements of the vector space $A/[A, A]$ give rise to *functions* on $Repr_n^A$. Namely, any $a \in A$ modulo commutators $[A, A]$ gives the function $Trace(\hat{a})$. Thus, we get a homomorphism of vector spaces

$$A/[A, A] \longrightarrow \mathcal{O}(Repr_n^A).$$

This has an obvious extension to the homomorphism of algebras

$$Sym^*(A/[A, A]) \longrightarrow \mathcal{O}(Repr_n^A).$$

We will denote the element of $\mathcal{O}(Repr_n^A)$ corresponding to $\phi \in Sym^*(A/[A, A])$ by $Trace(\phi)$.

In the following examples we can multiply our differential-geometric objects by an arbitrary element of the algebra $Sym^*(A/[A, A])$.

Vector fields Any derivation $\xi \in Der(A)$ of A gives a vector field $\tilde{\xi}$ on $Repr_n^A$.

Differential forms Any element of ΩA (see 1.1.1) gives a matrix-valued differential form on $Repr_n^A$, and thus a scalar-valued differential form after taking the trace in the matrix algebra. Again, using the vanishing of the trace of commutators, one can see that the map $\Omega A \longrightarrow \Omega^*(Repr_n^A)$ goes through the quotient space

$$\Omega A/[\Omega A, \Omega A]_{super} \longrightarrow \Omega^*(Repr_n^A),$$

where the commutator in superalgebra ΩA is understood in the super sense. Also, one can easily see that the differential d in Ω_A maps to the de Rham differential in the usual forms $\Omega^*(Repr_n^A)$. For example, the algebra $A = k \oplus k$ has a closed 2-form. The definition of "noncommutative differential forms" as elements of $\Omega A/[\Omega A, \Omega A]_{super}$, is due to M. Karoubi.

De Rham cohomology   It is natural in light of the previous discussion to define the de Rham cohomology of A as the cohomology of the complex $\Omega A/[\Omega A, \Omega A]$. Using results of Cuntz and Quillen [CQ2], one can show that $H_{dR}^n(A)$ defined in this way coincides with the $Z/2Z$-graded periodic cyclic homology $HP_n(A)$ for $n = 0, 1$, and with the reduced periodic cyclic homology

$$\widetilde{HP}_{n(mod\ 2)}(A) := HP_{n(mod\ 2)}(A)/HP_{n(mod\ 2)}(k)$$

for $n \geq 2$. In other words, $H_{dR}^n(A)$ coincides with $HP_{n(mod\ 2)}(A)$ for $n = 0, 1, 3, 5, \ldots$ and has one dimension less than $HP_{n(mod\ 2)}(A)$ for $n = 2, 4, 6, \ldots$.

Volume element, polyvector fields   It looks plausible from examples that one can define noncommutative analogues of polyvector fields and volume elements, but we would like to postpone the discussion of these notions. It seems that the divergence of a derivation of A with respect to a volume element belongs to the symmetric square of the vector space $A/[A, A]$. As an exercise to the reader we leave the following simple

Lemma   *If $A = k\langle x_1, \ldots x_d \rangle$, then there exists a linear map*

$$div : Der(A) \longrightarrow Sym^2(A/[A, A])$$

such that for any $\xi \in Der(A)$ the divergence of the corresponding vector field $\tilde{\xi}$ on $\mathbb{A}^{dn^2} = Repr_n^A$ with respect to the standard volume element, is equal to $Trace(div(\xi))$.

The last remark which we would like to make is that the correspondence (*Smooth algebras*) $\longrightarrow$ (*Representation schemes*) gives a "justification" of the notation TA in 1.1. Namely, there is a natural isomorphism between $Repr_n^{TA}$ and the total space of the tangent bundle $TRepr_n^A$.

1.3.3   Possible relations to matrix integrals and to M-theory

One can imagine that in the case $k = \mathbb{C}$, for a given "volume element" *vol* on A, a "function" $f \in Sym^*(A/[A, A])$ and for a the set of real points γ_n in $Hom(A, Mat(n \times n, \mathbb{C}))$ under some anti-holomorphic involution, one can take the integral of $exp(Trace(f)) \times vol$ and get a "matrix model", an infinite sequence of numbers

$$I_n := \int_{\gamma_n} e^{Trace(f)} vol$$

parameterized by the dimension $n = 1, 2, \ldots$. In mathematical physics such integrals have been extensively studied. Typically, one integrates over the space of hermitian $n \times n$ matrices, or over real, or unitary matrices. In multi-matrix models, the integration is taken over the set of d-tuples $(X_1, \ldots, X_d)$ of hermitian matrices. It is believed that the asymptotic behavior of I_n as $n \longrightarrow \infty$ is related to some kind of string theory.

Also, one recently proposed matrix model, the so-called M-theory one, is formulated in the same fashion. Roughly speaking, M-theory on the space-time manifold $X = \mathbb{R}^d$ is a matrix theory corresponding to the free algebra with d generators, the noncommutative affine space. M-theory on curved spaces should correspond to nontrivial smooth noncommutative algebras.

1.3.4 Double tangent space and formal noncommutative structure

Here we would like to give some examples of natural nonclassical structures on manifolds $Repr_n^A$. First of all, there is a natural vector bundle $T_{(2)}$ on the square $Repr_n^A \times_k Repr_n^A$, together with the identification of its pullback via the diagonal embedding $\Delta : Repr_n^A \longrightarrow Repr_n^A \times_k Repr_n^A$ with the tangent bundle $T_{Repr_n^A}$.

In what follows we will describe the bundle $T_{(2)}$. First of all, k-points of $Repr_n(A)$ can be identified with equivalence classes of pairs $(\mathcal{E}, pr_{\mathcal{E}})$ where $\mathcal{E}$ is an A-module finite-dimensional over k, and $pr_{\mathcal{E}} : A^n \longrightarrow \mathcal{E}$ is an epimorphism to $\mathcal{E}$ from the standard n-dimensional free A-module such that the set of n canonical generators of A^n maps to a basis of $\mathcal{E}$ over k. The fiber of $T_{(2)}$ at the pair $((\mathcal{E}, pr_{\mathcal{E}}), (\mathcal{F}, pr_{\mathcal{F}}))$ is defined as

$$Hom_{A-mod}(Ker(pr_{\mathcal{E}}), \mathcal{F}).$$

Let us prove that it is a finite-dimensional vector space. It is easy to see that it fits into an exact sequence

$$0 \longrightarrow Hom_{A-mod}(\mathcal{E}, \mathcal{F}) \longrightarrow \mathcal{F}^n \longrightarrow Hom_{A-mod}(Ker(pr_{\mathcal{E}}), \mathcal{F})$$
$$\longrightarrow Ext^1_{A-mod}(E, F) \longrightarrow 0.$$

All spaces except the one in question are finite-dimensional by the theorem in 1.1.3. Also, it is easy to see that on the diagonal $\mathcal{E} = \mathcal{F}$, $pr_{\mathcal{E}} = pr_{\mathcal{F}}$, the space $Hom_{A-mod}(Ker(pr_{\mathcal{E}}), \mathcal{F})$ coincides with the tangent space to $Repr_n^A$ at the point $(\mathcal{E}, pr_{\mathcal{E}})$.

Another structure on $Repr_n^A$ is the *formal noncommutative structure* in the sense of Kapranov. There is a canonical sheaf $\mathcal{O}_{Repr_n^A}^{noncomm}$ of formal noncommutative functions on $Repr_n^A$ with the quotient $\mathcal{O}_{Repr_n^A}$. For example, if $A = k\langle x_1, \ldots, x_d \rangle$ and $n = 1$ then the global sections of the sheaf $\mathcal{O}_{Repr_n^A}^{noncomm}$ is the projective limit $lim(A/I_n)$ where I_n, $n \geq 1$ is a decreasing sequence of two-sided ideals

$$I_n = \sum_{l \geq 1, (m_1, \ldots, m_l): \sum m_i = n} A \cdot A^{[m_1]} \cdot A \cdot A^{[m_2]} \cdot \ldots \cdot A^{[m_l]} \cdot A.$$

Here $A^{[m]}$ for $m \geq 1$ is defined as the linear span of the set of commutators of depth m:

$$[a_0, [a_1, [\ldots [a_{m-1}, a_m] \ldots] \in A, \quad a_0, a_1, \ldots, a_m \in A.$$

We refer the reader to [Ka] for the definition of the formal noncommutative structure and its differential-geometric meaning. We would like only to mention that Kapranov proposed a general construction of a formal non-commutative structure on moduli spaces of objects in abelian categories, like the category of coherent sheaves on algebraic varieties, etc. His construction admits a useful extension. Namely, if $\mathcal{C}$ is a k-linear triangulated category and $\mathcal{E}_0$ is a fixed object, then one can try to consider the "moduli space" of pairs $(\mathcal{E}, p)$, where $\mathcal{E}$ is an object of $\mathcal{C}$ and $p : \mathcal{E}_0 \longrightarrow \mathcal{E}$ is a morphism. The tangent complex for such a pair is $RHom_{\mathcal{C}}(\mathcal{E}, Cone(p))$. It carries a natural structure of an associative non-unital algebra, and the arguments from [Ka] are applicable there.

1.4   What we would like to do?

It is clear from the previous discussion that the subject of smooth noncommutative geometry merits further development. All our previous constructions and examples are *affine*, in particular they give affine representation schemes. It would be desirable to give a definition of "smooth noncommutative schemes", and also of "quasi-projective noncommutative schemes", such that it would contain finitely generated (or maybe finitely presented) smooth algebras as an affine case, and give rise to smooth schemes via a certain functor of "representations to $Mat(n \times n, k)$". We expect that representation schemes carry bitangent bundles, Kapranov's formal non-commutative structure, etc.

In the second part of this article, we describe a general approach to noncommutative algebraic geometry based on flat pre-topology. We

choose (temporarily) the name "spaces" (or "noncommutative spaces")
instead of "schemes" for several reasons. Noncommutative schemes were
introduced in the earlier works of one of us (see [R2]), where analogues
of Zariski topology were studied. Also, our formalism describes not only
usual schemes but algebraic spaces as well. As the reader will see, an
amazing variety of basic constructions in normal algebraic geometry can
be extended to the noncommutative setting. Our category of noncommutative spaces could be used in other situations. For example, quantum
projective spaces arising from Sklyanin algebras also fit into our definition.

In the third part we study one particular space which we call "noncommutative projective space" and denote by $N\mathbb{P}_k^n$. We believe that $N\mathbb{P}_k^n$
is one of the principal examples of what should be called a "smooth projective noncommutative variety".

2 Noncommutative spaces

2.1 Covers and refinements

Here we describe intermediate objects which are not yet spaces (there
are no structure sheaves on them). These objects are machines produc-
ing abelian categories. Essentially all definitions can be made in general
monoidal categories (even in nonadditive categories), instead of the category of vector spaces over k with the monoidal structure given by the
tensor product.

2.1.1 Category of covers

Definition Objects of category $Covers_k$ are given by the following

  DATA:
  1) an associative algebra $B \in Ob(Alg_k)$,
  2) $M \in Ob(B - mod - B)$, a bimodule over B,
  3) $m_M : M \longrightarrow M \otimes_B M$, a homomorphism of bimodules,
  4) $e_M : M \longrightarrow B$, also a homomorphism of bimodules,

satisfying the following

  AXIOMS:
  1) M is faithfully flat as the right B-module, i.e., the functor $M \otimes_B :$
$B - mod \longrightarrow B - mod$ is exact and does not kill any nonzero morphism,
  2) M with m_M and e_M is a coassociative coalgebra with counit in the
monoidal category of $B \otimes B^{opp}$-modules.

We will usually denote objects of the category $Covers_k$ as pairs (B, M)

skipping data m_M and e_M. Morally, we consider pairs (B, M) as "non-commutative stacks" together with a finite affine cover with affine pairwise intersections of members of the cover (see examples in the next subsection).

For every object (B, M) of $Covers_k$ we have an abelian category $QCoh(B, M)$ whose objects are pairs $(\mathcal{E}, m_\mathcal{E})$ where $\mathcal{E}$ is a B-module and $m_\mathcal{E} : \mathcal{E} \longrightarrow M \otimes_B \mathcal{E}$ is a homomorphism of B-modules which defines a coaction of coalgebra M on $\mathcal{E}$. We call objects of $QCoh(B, M)$ *quasi-coherent sheaves* on the "noncommutative stack" corresponding to (B, M). Presumably, one can justify the name "sheaf" introducing an appropriate Grothendieck topology.

Definition Morphisms f in the category $Covers_k$ from (B_1, M_1) to (B_2, M_2) are given by the following

   DATA:
   1) $f_B : B_2 \longrightarrow B_1$, a morphism of algebras,
   2) $f_M : M_2 \longrightarrow M_1$, a morphism of k-vector spaces
satisfying the following

   AXIOMS:
   1) the diagram where vertical arrows are structure morphisms of M_i as B_i-bimodules $(i = 1, 2)$ is commutative:

$$
\begin{array}{ccc}
B_2 \otimes M_2 \otimes B_2 & \xrightarrow{\ f_B \otimes f_M \otimes f_B\ } & B_1 \otimes M_1 \otimes B_1 \\
\downarrow & & \downarrow \\
M_2 & \xrightarrow{\quad f_M \quad} & M_1
\end{array}
$$

   2) and 3) two analogous diagrams including coproduct morphisms m_{M_i} and counit morphisms e_{M_i} respectively, are commutative.

Note that the direction of the morphism $f = (f_B, f_M)$ is *opposite* to the direction of the pullback morphisms of algebras and bimodules.

Every morphism $f = (f_B, f_M) : (B_1, M_1) \longrightarrow (B_2, M_2)$ of covers defines a functor $f^* : QCoh(B_2, M_2) \longrightarrow QCoh(B_1, M_1)$, the pullback of quasi-coherent sheaves. This functor maps an object $(\mathcal{E}, m_\mathcal{E})$ to the object $(B_1 \otimes_{B_2} \mathcal{E}, m'_\mathcal{E})$ where the coaction morphism

$$
m'_\mathcal{E} : B_1 \otimes_{B_2} \mathcal{E} \longrightarrow M_1 \otimes_{B_1} B_1 \otimes_{B_2} \mathcal{E} = M_1 \otimes_{B_2} \mathcal{E}
$$

is defined in a natural way using f_B, f_M and $m_\mathcal{E}$. One can show that the function f^* has a right adjoint function f_*.

2.1.2 Examples of covers

FC1) Let $S/Spec(k)$ be a separated quasi-compact scheme, e.g., a quasi-projective scheme. We choose a finite cover $(\mathcal{U}_i)_{i \in I}$ of S in the Zariski topology by affine schemes. It follows from separatedness that pairwise intersections $\mathcal{U}_i \cap \mathcal{U}_j$ are again affine. The algebra B and the bimodule M are defined as

$$B := \mathcal{O}(\sqcup_i \mathcal{U}_i), \quad M := \mathcal{O}(\sqcup_{i,j}(\mathcal{U}_i \cap \mathcal{U}_j))$$

and the structure of the coalgebra on M is the natural one. It follows from the usual descent theory that the category $QCoh(B, M)$ is equivalent to the category of quasi-coherent sheaves on S.

FC1') The same statement holds for affine covers of separated quasi-compact schemes in fpqc topology, for algebraic stacks, etc.

FC2) Let A be an associative algebra and $M := A$ with the natural structure of an A-bimodule, and a coalgebra in the category A-mod-A. Then $QCoh(A, M)$ is equivalent via the tautological functor to the category A-mod.

FC3) Let A be an associative algebra, and let $B \supset A$ be a larger algebra such that B is faithfully flat as the right A-module. Then we define M as $B \otimes_A B$ with the natural structure of a coalgebra in the category B-mod-B. We have a natural morphism of covers

$$f : (A, A) \longrightarrow (B, M)$$

Theorem *The morphism f as above defines an equivalence f^* between the categories $QCoh(B, M)$ and $QCoh(A, A) = A$-mod.*

This theorem follows from the general Barr–Beck theorem. Recall that a *comonad* (or *cotriple*) in the category C consists of a functor $T : C \to C$ and two functor morphisms $\delta : T \to T^2$ and $\epsilon : T \to Id_C$ such that $T\delta \circ \delta = \delta T \circ \delta$ and $\epsilon T \circ \delta = id_T = T\epsilon \circ \delta$. A *coaction* of the comonad T is a morphism $c : X \to T(X)$ such that $Tc \circ c = \delta(X) \circ c$ and $\epsilon(X) \circ c = id_X$. We denote by T-$comod$ the category of T-*comodules*, i.e., objects of C endowed with a coaction of T, with naturally defined morphisms: a morphism $(X, c) \to (X', c')$ is a morphism $f : X \to X'$ such that $Tf \circ c = c' \circ f$. An arbitrary pair of adjoint functors $F : C_1 \longrightarrow C_2$, $G : C_2 \longrightarrow C_1$ with adjunction morhisms $\epsilon : FG \to Id_{C_2}$ and $\eta : Id_{C_1} \to GF$ determines a comonad (T, δ, ϵ), where $T = FG$, $\delta = F\eta G$, and there is an obvious functor $C_1 \longrightarrow T$-$comod$, $Y \mapsto (F(Y), F\eta(Y))$. The following version of Barr–Beck's theorem is sufficient for our needs.

Theorem   *Let $F : C_1 \longrightarrow C_2$ be a functor between two categories having a right adjoint functor $G : C_2 \longrightarrow C_1$. Assume that the following conditions hold:*

(a) *a pair of arrows $f, g : X \rightrightarrows Y$ has a kernel if its image by F has a kernel;*

(b) *if $h : Z \longrightarrow X$ is such that $f \circ h = g \circ h$ and Fh is the kernel of the pair (Ff, Fg), then h is a kernel of (f, g).*

Then the canonical functor $C_1 \longrightarrow T$-comod is an equivalence of categories.

We apply this theorem to the case $C_1 = A\text{-mod}$, $C_2 = B\text{-mod}$ and $F = B \otimes_A$.

Note that the arbitrary pairs of morphisms in Barr–Beck's theorem can be replaced by so-called *coreflexive pairs*, i.e., the pairs $f, g : X \rightrightarrows Y$ such that there exists a morphism $e : Y \to X$ such that $e \circ f = id_X = e \circ g$. And even this condition can be weakened (cf. [MLM], IV.4).

2.1.3   Refinements of covers

Let (B_1, M_1) be a cover and let $i : B_1 \longrightarrow B_2$ be an inclusion of algebras such that B_2 is faithfully flat as a right B_1-module. We define M_2 as $B_2 \otimes_{B_1} M_1 \otimes_{B_1} B_2$ with the natural structure of a coalgebra in the category $B_2\text{-mod-}B_2$. It is easy to see that (B_2, M_2) is again a cover and we have a natural morphism of covers $f : (B_2, M_2) \longrightarrow (B_1, M_1)$.

Definition   A morphism of covers isomorphic to f as above is called a refinement morphism.

We denote by Ref the class of refinement morphisms. Analogous to the example FC3) from the previous subsection, the functors f^* for $f \in Ref$ are equivalences of categories of quasi-coherent sheaves. Here we apply the Barr–Beck theorem to the functor which acts from $QCoh(B_1, M_1)$ to $B_2\text{-mod}$ and is given by the formula $F(\mathcal{E}, m_{\mathcal{E}}) = B_2 \otimes_{B_1} \mathcal{E}$.

The class of refinements is closed under compositions. Also, any pullback of a refinement morphism by a refinement morphism is again a refinement morphism. This follows from a theorem of P. M. Cohn (See [C]).

2.2   Noncommutative spaces

In this section we give the definition of noncommutative spaces (in several steps). Morally one should think about the noncommutative space X

as about an abelian category $QCoh(X)$ and two adjoint functors $\pi_* : QCoh(X) \longrightarrow k\text{-}mod$, $\pi^* : k\text{-}mod \longrightarrow QCoh(X)$. The pullback $\pi^*(k^1)$ of the standard 1-dimensional k-module is the "structure sheaf" $\mathcal{O}_X$.

2.2.1 Covers of noncommutative spaces

We define covers of noncommutative spaces adding some data and an axiom to the definition of the category of covers:

Definition Objects of category $Covers_k^{sp}$ are pairs (C, s_C) where $C = (B, M)$ is a cover and $s_C : B \otimes B \longrightarrow M$ a homomorphism of bimodules which is an epimorphism and also a morphism of coalgebras. Morphisms in $Covers_k^{sp}$ are morphisms of covers compatible with structure epimorphisms from $B \otimes B$.

In other words, we demand that M be a cyclic bimodule generated by an element $m := s_C(1_B \otimes 1_B)$ which behaves well with respect to the coalgebra structure. Such an M is given by a left ideal $Ker(s_C)$ in $B \otimes B^{opp}$ satisfying a complicated system of axioms.

We will call objects of $Covers_k^{sp}$ space covers. Notice that all our examples of covers (except stacks) are automatically space covers.

The morphism s_C gives rise to a canonical morphism from (C, s_C) to the refinement $(B, B \otimes B)$ of the "point" object $Spec(k) := ((k, k), id)$ (the final object in $Covers_k^{sp}$). We call the pullback of the standard 1-dimensional module k^1 under this morphism the structure sheaf $\mathcal{O}_C$. It is represented by the 1-dimensional free B-module $B^1 = B$ with the coaction of M arising from s_C. Also, for any refinement $f : (B_2, M_2) \longrightarrow (B_1, M_1)$ from cover to a space cover there exists a unique structure of a space cover on (B_2, M_2) such that f becomes a a morphism of space covers.

We define *equivalence classes of noncommutative spaces* as equivalence classes of space covers modulo the relation generated by refinements. In order to define the "right" *category* of noncommutative spaces, we need to have some additional work.

2.2.2 The category of noncommutative spaces

Definition The class of objects in the category $Spaces_k$ of noncommutative spaces over k is defined as the class of space covers. The set of morphisms in $Spaces_k$ from (C, s_C) to $(C', s_{C'})$ is defined as the set of equivalence classes of pairs $f = (f^*, iso_f)$ where $f^* : QCoh(C') \longrightarrow QCoh(C)$ is a functor which has the right adjoint, and $iso_f : \mathcal{O}_C \longrightarrow f^*(\mathcal{O}_{C'})$ is an isomorphism in $QCoh(C)$.

We call morphisms of the category $Spaces_k$ "maps". One can show that the automorphism group of any representative (f^*, iso_f) of a map is trivial. We denote the right adjoint to f^* by f_*. Thus, one can canonically associate functors f_* and f^* to maps.

For any cover $C = (B, M)$, we denote by $\pi_C : (B, B) \longrightarrow (B, M)$ the obvious morphism of covers. It can be shown that for any map f from $((B, M), s_C)$ to $((B', M'), s_{C'})$, the functor $(\pi_{C'})_* \circ f_* \circ (\pi_C)^* : B\text{-}mod \longrightarrow B'\text{-}mod$ is given by tensoring over B by a bimodule $F \in B'\text{-}mod\text{-}B$. Moreover, this bimodule is cyclic with a given generator, i.e., it is equal to $(B' \otimes B^{opp})/I_f$ where $I_f \subset B' \otimes B^{opp}$ is a left ideal. Conversely, f can be canonically reconstructed from I_f.

Theorem *The category of separated quasi-compact schemes over k is equivalent to a full subcategory of $Spaces_k$. The category Alg_k^{opp} is also equivalent to a full subcategory of $Spaces_k$.*

Theorem *Construction $(B, M) \longrightarrow QCoh(B, M)$ extends to a functor from the category $Spaces_k$ to the category of abelian k-linear categories. For separated quasi-compact schemes, this gives the usual quasi-coherent sheaves of commutative algebraic geometry. For associative algebras, this gives categories of left modules.*

One can consider noncommutative stacks by eliminating the condition that the structure homomorphism $s_S : B \otimes B \longrightarrow M$ be surjective. We will not develop here the more general theory of stacks.

2.3   Affine covers

We call *affine spaces* noncommutative spaces corresponding to the cover of the type (B, B) where B is an algebra. Abusing notations, we will denote the space (B, B) by $Spec(B)$, although here we do not use points of the spectrum ([R3], [R1], [R4]) at all. The algebra B can be identified with

$$Hom_{QCoh(Spec(B))}(\mathcal{O}_{Spec(B)}, \mathcal{O}_{Spec(B)}) \ .$$

Also, B coincides as a set with the set of maps from (B, B) to the (non)-commutative affine line $\mathbb{A}_k^1 = N\mathbb{A}_k^1 := Spec(k[t])$. In general, the d-dimensional noncommutative affine space $N\mathbb{A}_k^d$ is defined as $Spec(k\langle x_1, \ldots, x_d \rangle)$. The algebra structure on $\mathcal{O}(S) = Maps(S, N\mathbb{A}_k^1)$ is induced by the algebra structure on $N\mathbb{A}_k^1$ considered as an object of $Spaces_k$.

Let $\pi : Spec(B) \longrightarrow S$ be a cover of a space by an affine space. We can take the pullback of cover π by itself, and then repeat the procedure. In this way we form an infinite sequence of affine spaces $Spec(B^{(n)})$, $n \geq 1$ covering S, with the initial condition $B^{(1)} = B$. The algebra of functions on $S^{(n)}$ is generated by n copies $i_1(B), \ldots, i_n(B)$ of the algebra B subject to the set of relations which we will describe now.

Relations *Let $z = \sum_{\alpha} a_{\alpha} \otimes b_{\alpha} \in B \otimes B$ be an arbitrary element of the kernel of the structure map $s_S : B \otimes B \longrightarrow M$. Then for any $k \neq, l$, $1 \leq k, l \leq n$, the element $\sum_{\alpha} i_k(a_{\alpha}) \cdot i_l(a_{\alpha})$ is equal to zero in $B^{(n)}$.*

Also, it is easy to see that the correspondence $n \mapsto Spec(B^{(n)})$ extends naturally to a contravariant functor from the category of nonempty finite sets to the category of affine spaces. We will need this description of algebras $B^{(n)}$ in 3.2 for the justification of a definition of the noncommutative projective space.

2.4 Cohomology of quasi-coherent sheaves

As we mentioned in the previous section, for every space S, there exists a distinguished object $\mathcal{O}_S$ in the category $QCoh(S)$. It is easy to see that the functor $\mathcal{E} \mapsto Hom(\mathcal{O}_S, \mathcal{E})$ is the right adjoint to the inverse image functor $f_S^*; QCoh(point) \longrightarrow QCoh(S)$ for the canonical map $f_S : S \longrightarrow point = Spec(k)$. We denote the vector space $Hom(\mathcal{O}_S, \mathcal{E})$ as $\Gamma(\mathcal{E})$ (the space of global sections of $\mathcal{E}$). The functor Γ is left exact.

We would like to define now the derived functor for Γ. First of all, it makes sense because of the following

Theorem *Category $QCoh(S)$ has enough injective objects.*

The proof is the following. Let us chose an affine cover $\pi : Spec(B) \longrightarrow S$. The functor π^* has a right adjoint π_*. The category B-*mod* (canonically $\simeq QCoh(Spec(B)))$ has enough injective objects. Now, for any quasi-coherent sheaf $\mathcal{E}$ on S, let us chose an embedding of $\pi^* \mathcal{E}$ into an injective B-module I. Due to the fact that the inverse image functor π^* is exact, $\pi_* I$ is again an injective object. Since the functor π^* is faithful, the natural homomorphisms $\mathcal{E} \longrightarrow \pi_* \pi^* \mathcal{E}$ is a monomorphism. And the image $\pi_* \pi^* \mathcal{E} \longrightarrow \pi_* I$ of the embedding $\pi^* \mathcal{E} \longrightarrow I$ is a monomorphism too, because the direct image functor π_* is left exact (as any functor having a left adjoint).

Thus, one can proceed and define the derived functor $R\Gamma$ (and more generally $RHom$, more generally, the derived functor of any left exact functor) using injective resolutions.

Theorem *For any cover* (B, M) *of* S *and for any object* $(\mathcal{E}, m_{\mathcal{E}}) \in$ $QCoh(B, M)$, *the cohomology of* S *with coefficients in* $\mathcal{E}$ *can be calculated via the Čech complex*

$$\mathcal{E} \longrightarrow M \otimes_B \mathcal{E} \longrightarrow M \otimes_B M \otimes_B \mathcal{E} \longrightarrow \dots$$

One can extend the previous result to the relative case. For every morphism $f : (C_1, s_{C_1}) \longrightarrow (C_2, s_{C_2})$ in category $Covers_k^{sp}$, the functor f^* admits a right adjoint f_*. Since *the direct image functor* f_* *is left exact,* there is a derived functor for f_*.

2.5   Definition of coherent sheaves

In commutative algebraic geometry, the abelian category of coherent modules are usually considered in the case of noetherian rings. In the noncommutative setting, our principal examples are free algebras in several indeterminates which are far from being noetherian. Nevertheless, one can define a reasonable abelian category of modules over smooth algebras as well.

Definition A module M over an algebra A is called coherent iff it is finitely presented, i.e., there exists an exact sequence

$$F_{-1} \longrightarrow F_0 \longrightarrow M \longrightarrow 0,$$

where F_{-1}, F_0 are free finitely generated A-modules.

We denote $Coh(A)$ the full subcategory of $A - mod.$ objects of which are coherent modules.

Lemma *For any hereditary algebra* A *(e.g., for a formally smooth* A), *the category* $Coh(A)$ *is an abelian category.*

It is enough to check that the kernel of any morphism $\phi : M \longrightarrow N$ of coherent A-modules is coherent. The proof is the following. In hereditary abelian categories any object in the bounded derived category is isomorphic to its cohomology. Thus, $Ker(\phi)$ is a direct summand in $D^b(A-mod)$ of a perfect complex. Using telescope construction, we see that $Ker(\phi)$ is quasi-isomorphic in an infinite complex bounded from above of finitely generated projective modules. From this it follows that $Ker(\phi)$ is coherent.

The property of a module over arbitrary algebra to be (or not to be) finitely presented is preserved under faithfully flat extensions of the algebra. This implies that for spaces, one can define the notion of a finitely presented quasi-coherent sheaf passing to an arbitrary affine cover. The property of being finitely presented is independent of the cover.

For any noncommutative space S admitting a cover by a smooth or noetherian affine space, the category of finitely-presented quasi-coherent sheaves is abelian. In such a case we call this the category $Coh(S)$ of coherent sheaves.

2.6 Properties of morphisms

First of all, affine and flat morphisms are defined using appropriate properties of functors between abelian categories of quasi-coherent sheaves.

Definitions *Let $f : X \to Y$ be a morphism. We call f affine iff f_* is faithful and has a right adjoint; we call f flat iff f^* is exact, and faithfully flat iff f^* is exact and faithful.*

Notice that for any morphism f, the functor f_* is left exact, and the functor f_* is right exact. Also, for affine f the functor f_* is exact.

We call an affine scheme morphism $Spec(A) \to Spec(B)$ a *thickening* if the corresponding algebra morphism $B \to A$ is an epimorphism and its kernel is a nilpotent ideal.

Following the pattern of commutative algebraic geometry [EGA, IV.17], we give the following

Definitions *1) Let $f : X \to Y$ be a morphism. We call f formally smooth (resp. formally nonramified, resp. formally étale) if, for any affine scheme $Spec(A)$, any thickening $Spec(T) \to Spec(A)$, and any morphism $Spec(A) \to Y$, the canonical morphism*

$$Hom_Y(Spec(A), X) \to Hom_Y(Spec(T), X)$$

defined by the immersion $Spec(T) \to Spec(A)$ is surjective (resp. injective, resp. bijective).

2) A k-space X will be called formally smooth if the canonical morphism $X \to Spec(k)$ is formally smooth.

Remarks (i) If X is affine, then the definition of the formal smoothness given here is in accordance with the one in 1.1.

(ii) To check that f is formally smooth (resp. formally nonramified, resp. formally étale), it suffices to do it in the case when the square of the kernel J of the algebra epimorphism $A \to T$ is zero.

(iii) The properties of f defined in the second group of definitions in
2.6 are those represented by the f functor $Y' \mapsto Hom_Y(Y', X)$ from the
category dual to the category of Y-spaces to the category *Sets*. They
make sense for any functor $(Spaces_k/Y)^{op} \to Sets$, representable or not.

(iv) It follows from the definition that f is formally étale iff it is both
formally smooth and formally nonramified. One can show that $Spec(A)$
is étale (i.e. $Spec(A) \to Spec(k)$ is étale) iff the algebra A is *separable*.
The latter means that A is a projective A-bimodule, or equivalently, A
has dimension zero with respect to Hochschild cohomology. An example
of a separable algebra is the algebra $Mat(n \times n, k)$ of $n \times n$ matrices.

3   Noncommutative projective space

3.1   Imitation of the definition à la Grothendieck

For any integer $d \geq 1$, we can try to define the *noncommutative projective
space* $N\mathbb{P}_k^{d-1}$ as a space representing a certain contravariant functor from
the category of affine spaces to the category *Sets*.

Let A be an algebra. The set $Map(Spec(A), N\mathbb{P}_k^{d-1})$ should be func-
torially identified with the set of quotient modules $\mathcal{F}$ of the standard free
d-dimensional A-module A^d, such that *locally* $\mathcal{F}$ is isomorphic to the free
1-dimensional A-module. The last condition should imply that $\mathcal{F}$ is projective.

The problem with this definition is the word "locally". We did not yet
introduce any topology on noncommutative spaces, even on affine ones.
One cannot use covers as defined as faithfully flat morphisms because this
class is not closed under pullbacks by arbitrary morphisms. Nevertheless,
it seems that one can give a definition of $N\mathbb{P}_k^{d-1}$ which is sufficiently robust
and works for a class of topologies satisfying some soft conditions. We will
not present here explicitly these conditions. The key to the "definition"
of $N\mathbb{P}_k^{d-1}$ is a calculation of algebras $B^{(n)}$ in 2.3.

3.2   Explicit cover

We will construct a cover of $N\mathbb{P}_k^{d-1}$ analogous to Jouanolou's cover of the
projective space in commutative algebraic geometry. Recall that this is a
cover by the affine quadric

$$\{(x_1, \ldots, x_d; y_1, \ldots, y_d) | \sum_{i=1}^{d} x_i y_i = 1\} \longrightarrow \{(x_1 : \ldots : x_d)\} = \mathbb{P}^d(k) \ .$$

Let B be the associative algebra generated by $2d$ variables $x_1, \ldots, x_d, y_1, \ldots, y_d$ satisfying the relation $\sum_{i=1}^{d} y_i x_i = 1$. The space $Spec(B)$ represents the following functor on affine spaces: $Maps(Spec(A), Spec(B))$ is the set of epimorphisms $p : A^d \longrightarrow \mathcal{F}$ of A-modules together with an isomorphism $i : A^1 \simeq \mathcal{F}$ of A-modules, and a splitting s of A^n into the direct sum of $\mathcal{F}$ and a complementary module $\mathcal{F}'$. Also, for each $n \geq 1$, we can consider the functor whose value on $Spec(A)$ is the set of collections $(\mathcal{F}, p; i_1, s_1, \ldots, i_n, s_n))$, where $\mathcal{F}$ is a quotient module of A^d, $p : A^d \longrightarrow \mathcal{F}$ is an epimorphism, and for each $k = 1, \ldots, n$, the quadruple $(\mathcal{F}, p; i_k, s_k)$ belongs to $Maps(Spec(A), Spec(B))$. It is easy to see that this functor is representable by an affine space whose algebra of functions we denote by $B^{(n)}$.

Thus we have constructed a contravariant functor from the category of nonempty finite sets to the category of affine spaces. One can check by direct calculations that the algebra $B^{(n)}$ is generated by n copies $i_1(B), \ldots, i_n(B)$ modulo relations of the type described in 2.3. The bimodule M in this case is $(B \otimes B)/I$ where I is the left $B \otimes B^{opp}$-ideal generated by the following set of elements:

$$e_j := -1 \otimes x_j + \sum_{i=1}^{d} x_j y_i \otimes x_i, \quad j = 1, \ldots, d .$$

Definition The noncommutative projective space $N\mathbb{P}_k^{d-1}$ is defined as the cover (B, M) where B and $M = (B \otimes B)/I$ are as above.

Theorem *The bimodule M has a canonical structure of a coalgebra in the category of bimodules induced by the epimorphism $B \otimes B \longrightarrow M$; it is faithfully flat as the right B-module. The functor from the category of nonempty finite sets to the category of affine spaces associated with the cover (B, M), as in 2.3, is canonically isomorphic to the functor constructed here in 3.2.*

Analogously, one can imitate the Grothendieck definition of projective space for coherent sheaves and introduce the relative projective space $\mathbb{P}(\mathcal{E})$ for any noncommutative space S and a finitely generated quasi-coherent sheaf $\mathcal{E}$ on S.

3.3 Derived category of quasi-coherent sheaves

Here we will only state the results without proofs (proofs will appear among other things in [KR]). We denote by Q_d the quiver, which has two vertices $\{v_0, v_1\}$ and d oriented edges all going from v_0 to v_1.

Theorem *For any $d \geq 1$ the category $QCoh(N\mathbb{P}_k^{d-1})$ has cohomological dimension 1. The bounded derived category $D^b(QCoh((N\mathbb{P}_k^{d-1}))$ is equivalent to the bounded derived category of representations of the quiver Q_d.*

Theorem *The algebra B described in 3.2 and the projective space $N\mathbb{P}_k^{d-1}$ are formally smooth. The bounded derived category of coherent sheaves on $N\mathbb{P}_k^{d-1}$ is of finite type (i.e., for any two objects $\mathcal{E}, \mathcal{F}$ we have $\sum_i rk(RHom^i(\mathcal{E}, \mathcal{F})) < +\infty$), and it is equivalent to the bounded derived category of finite-dimensional representations of Q_d. The group $K_0(Coh(N\mathbb{P}_k^{d-1}))$ is free abelian group with 2 generators.*

The main ingredient in the proof of both theorems is a noncommutative analogue of Beilinson's resolution of the sheaf of functions on the diagonal in $\mathbb{P}_k^{d-1} \times \mathbb{P}_k^{d-1}$. The noncommutative resolution is much shorter than the commutative one, and has length 2. There are two coherent sheaves $\mathcal{O}$ and $\mathcal{O}(1)$ (the universal quotient module $\mathcal{F}$ from 3.1) which generate in an appropriate sense the whole derived category on $N\mathbb{P}_k^{d-1}$.

There are two surprising particular cases: $d = 1$ for which the noncommutative projective space is more complicated than the usual one (which is a point), and the case $d = 2$. It is well-known that the category $D^b(Coh(\mathbb{P}_k^1))$ is equivalent to $D^b(Q_2 - mod_{finite})$. Thus, in the case $d = 2$, we have three different abelian categories with the same derived category: the noncommutative projective line, the commutative projective line, and the quiver Q_2.

3.4 Grassmannians

One can define Grassmannians $NGr_k(d', d)$ for $d', d \geq 1$ in the same fashion as the projective space, changing the requirement that the quotient module $\mathcal{E}$ of $\mathcal{O}^d$ be locally isomorphic to a free d'-dimensional module instead of to a free 1-dimensional module. We claim that in contrast to the commutative case we do not get a new space. It follows from the fact that there exists a *nonzero* algebra A such that A^1 is isomorphic A^2 (and thus to $A^{d'}$) as A-module. For example, one can take A equal to $End(V)$ where V is an infinite-dimensional vector space. On the faithfully flat extension which is obtained by taking the free product with such an algebra A, we identify conditions on $\mathcal{E}$ for all Grassmannians. Thus, $NGr_k(d', d)$ coincides with $N\mathbb{P}_k^{d-1} = NGr_k(1, d)$.

We expect that various remarkable identities in noncommutative linear algebra in the free skew field discovered by Gelfand and Retakh (see [GR]) can be interpreted as identities between morphisms of coherent sheaves on $N\mathbb{P}_k^{d-1}$ and on similar spaces.

3.5   Representation spaces

For a noncommutative space S, we can try to define representation schemes $Repr_n^S$. At least, it is clear how sets of k-points should look:

$$Reps_n^S(k) = Maps(Spec(Mat(n \times n, k)), S) \ .$$

It follows from the definition of the projective space in 3.1 and from the discussion in 3.4 that the set of k-points of $N\mathbb{P}_k^{d-1}$ is equal to the set of *all* nonzero quotient spaces of k^d. We leave to the reader as an exercise the description of maps from $Spec(Mat(n \times n, k))$ to $N\mathbb{P}_k^{d-1}$ for the case $n \geq 2$.

Let us express again our hopes: representation schemes $Repr_n^S$ for (formally) smooth nonaffine spaces S should carry all structures described in Part 1 (the double tangent bundle, Kapranov's formal noncommutative functions, etc.).

References

[C]   P. M. Cohn, On the free product of associative rings, Math. Z., v.71, 1959, 380–398

[CQ1]   J. Cuntz and D. Quillen, Algebra extensions and nonsingularity, *Journal of AMS* **8**, no. 2 (1995), 251–289.

[CQ2]   J. Cuntz and D. Quillen, Cyclic homology and nonsingularity, *Journal of AMS*, **8**, no. 2 (1995), 373–442.

[EGA]   A. Grothendieck, J.A. Dieudonné, Eléments de Géométrie Algébrique, Springer Verlag, New York - Heidelberg - Berlin, 1971.

[GR]   I. Gelfand and V. Retakh, Quasideterminants, I, *Selecta Math.* **3** (1997), 517–546.

[GZ]   P. Gabriel and M. Zisman, *Calculus of fractions and homotopy theory*, Springer Verlag, Heidelberg-New York, 1967

[K]   M. Kontsevich, Formal non-commutative symplectic geometry, in *The Gelfand Mathematical Seminars, 1990-1992*, Birkhäuser, Boston-Basel-Berlin (1993), pp. 173–187.

[Ka]   M. Kapranov, Noncommutative geometry based on commutator expansions, Math.AG/9802041 (1998), 48 pp.

[Kn]   D. Knutson, *Algebraic Spaces*, LNM 203, Springer-Verlag, 1971.

[KR]   M. Kontsevich, A. Rosenberg, Noncommutative separated spaces, in preparation.

[ML]   S. MacLane, *Categories for the Working Mathematicians*, Springer-Verlag, New York-Heidelberg-Berlin (1971).

[MLM] S. MacLane and I. Moerdijk, *Sheaves in Geometry and Logic*, Springer-Verlag, New York-Heidelberg-Berlin-London-Paris (1992).

[R1]   A.L. Rosenberg, *Noncommutative Algebraic Geometry and Representations of Quantized Algebras*, Kluwer Academic Publishers, Mathematics and its Applications, v.330 (1995), 328 pages.

[R2]   A.L. Rosenberg, Noncommutative schemes, *Compositio Mathematica* **112** (1998), 93–125.

[R3]   A.L. Rosenberg, Noncommutative local algebra, *Geometric and Functional Analysis (GAFA)* **4**, no.5 (1994), 545–585.

[R4]   A.L. Rosenberg, The spectrum of abelian categories and reconstruction of schemes, in *Algebraic and Geometric Methods in Ring Theory*, Marcel Dekker, Inc., New York, (1998), pp. 255–274.

Maxim Kontsevich   Alexander Rosenberg
I.H.E.S.   Department of Mathematics
35 Route de Chartres   Kansas State University
F-91440 Bures-sur-Yvette   137 Cardwell Hall
France   Manhattan, Kansas 66506
e-mail:maxim@ihes.fr   e-mail:rosenber@math.ksu.edu

AMS Subject Classifications: Primary 17Bxx; Secondary 16Sxx, 22E47

A Cycle for Integration Yielding the Zonal Spherical Function of Type A_n

A. Kazarnovski-Krol

Abstract

The integral of a certain multivalued form over a cycle Δ provides the zonal spherical function of type A_n. This paper is devoted to a quantum group analysis and verification of monodromy properties of the distinguished cycle Δ. The zonal spherical function in the case of the root system of type A_n is a particular conformal block of the WA_n-algebra.

0. Introduction

Let G be a connected real semisimple Lie group with finite center, K its maximal compact subgroup, G/K a complete Riemannian symmetric space. Let $T_g = T_g^\lambda, g \in G$, be a continuous unitary representation of G acting in a Hilbert space H, which contains a spherical vector ξ, i.e., $K\xi = \xi$ and assume that $(\xi, \xi) = 1$. Here λ is a parameter defining this representation. Then the function $\phi_\lambda(g) = (T_g^\lambda \xi, \xi)$ is called the zonal spherical function [26]. In particular, $\phi_\lambda(e) = 1$ and it is right and left K-invariant. The zonal spherical function is a common eigenfunction of the Laplace-Casimir operators:

$$\mathcal{L}\phi_\lambda(g) = \gamma(\mathcal{L})(i\lambda)\phi_\lambda(g)$$

where $\gamma(\mathcal{L})(i\lambda)$ is a homomorphism of the Laplace-Casimir operators into complex numbers [26,36]. Using the Cartan decomposition $G = KAK$, the zonal spherical function $\phi_\lambda(g)$ is considered to be a function on A. Let also $\mathfrak{a} = Lie(A)$; then $\lambda \in \mathfrak{a}^*$.

Restriction of the zonal spherical function to A is a common eigenfunction of the radial parts $\overset{\circ}{\mathcal{L}}$ of the Laplace-Casimir operators $\mathcal{L}$:

$$\overset{\circ}{\mathcal{L}}(\phi_\lambda(a)) = \gamma(\overset{\circ}{\mathcal{L}})(i\lambda)\phi_\lambda(a) \tag{1}$$

where γ is the Harish-Chandra homomorphism $a \in A$. Among others, the operator of second order plays the predominant role:

$$\overset{\circ}{\mathcal{L}}_2 = H_1^2 + \ldots + H_l^2 + \sum_{\alpha \in \mathcal{R}_+} m_\alpha \frac{e^\alpha + e^{-\alpha}}{e^\alpha - e^{-\alpha}} H_\alpha,$$

where m_α is a multiplicity of a restricted root α [36]. It turns out that one can consider m_α as complex parameters, but the condition $m_\alpha = m_{w\alpha}$ is required. So there are as many independent parameters as orbits of the Weyl group in the restricted root system. One may start with a second order differential operator with generic parameters, then recover the whole system of differential operators (radial parts)[37]. In the case of the root system of type A_n, this system was introduced by J. Sekiguchi [69], see also [54]. This system turns out to be holonomic; locally it has a $|W|$-dimensional space of solutions, where $|W|$ is the cardinality of the Weyl group W, cf. Corollary 3.9 of [37]. Among those solutions there is a distinguished one which corresponds to the zonal spherical function. It is characterized by analyticity (single-valuedness) at unity, cf. Theorem 6.9 [37].

We restrict ourselves to the case of the root sytem of type A_n. For the second order differential operator we use

$$\overset{\circ}{\mathcal{L}}_2 = \sum_i \left(z_i \frac{\partial}{\partial z_i} \right)^2 - k \sum_{i<j} \frac{z_j + z_i}{z_j - z_i} \left(z_i \frac{\partial}{\partial z_i} - z_j \frac{\partial}{\partial z_j} \right).$$

In [44] we provided an integral representation for the solutions of system (1); in [45] we described cycle Δ for integration yielding the zonal spherical function, obtained an explicit version of the Harish-Chandra expansion, and indicated that a basis of the solutions to the Sekiguchi-Debiard system (1) is provided by the conformal blocks of the WA_n-algebra.

This paper is devoted to a quantum group analysis and verification of monodromy properties of the distinguished cycle Δ (cycle Δ is recalled in 2.1 below). Cycle Δ serves as a contour for integration for the zonal spherical function of type A_n of a suitable multivalued form. The form is of a type considered in refs. [68,75] and thus the technique of quantum groups and R-matrices can be applied. We obtained the form using the very simple principle: there is only one line which passes through two given points. The zonal spherical functions for $SL(n+1, \mathbb{C})$ were calculated by I.M. Gelfand and M. Naimark [27] and using the same method by K. Aomoto for $SL(n+1, \mathbb{R})$ [4]. This provides us with the form for parameters $k = 1$ and $k = \frac{1}{2}$, where k is a half multiplicity of restricted roots. Now we use the principle and extend powers of factors linearly in k.

The obtained form has several advantages: cycle for integration yielding the zonal spherical function is real and compact; the number of variables of integration is independent of the parameter λ (manifestation of the flag manifold); there is no "complicated meromorphic factor." The integral of a certain multivalued form over cycle Δ provides the zonal spherical function of the type A_n, i.e., a particular solution of the Sekiguchi-Debiard sytem (1) which is analytic at unity. For certain integrality conditions on the spectral parameter λ, the zonal spherical function as a hypergeometric solution terminates and becomes the Jack polynomial.

Note: in refs. [56,16] the system (1) is proved to be related to a particular case of the trigonometric Knizhnik-Zamolodchikov equation. In particular, this implies that solutions to the system (1) can be obtained from the solutions of Knizhnik-Zamolodchikov equations by a symmetrization procedure (summation over the fiber; fiber is naturally parametrized by elements of the Weyl group W). Solutions to the Knizhnik-Zamolodchikov equations are given by certain multidimensional integrals, whose integrand has the standard part times a complicated meromorphic factor. This complicated meromorphic factor becomes even more complicated after symmetrization (summation over the fiber). We would like to emphasize that in this particular case, this unpleasant meromorphic factor is not needed, cf. [44], Theorem 6.3.

Knizhnik-Zamolodchikov equations originate in WZNW theory. Reduction from WZNW to W-algebras is well discussed in the literature [7,19] (quantum Drinfeld-Sokolov reduction). Results of [44] (absence of "meromorphic factor") imply that in the case of the root system of type A_n, solutions to the Sekiguchi-Debiard system (1) (which is isomorphic to Calogero-Sutherland model) are provided by certain conformal blocks of the WA_n-algebra. For more details about this approach we refer to our paper [46].

The quantum group approach assumes the following. With the multivalued form one associates the tensor product of irreducible highest weight modules over the quantum group. Nonzero elements of an antiinvariant part of homology of a certain type of discriminantal configuration correspond to the singular (vacuum) vectors of the tensor product of irreducible highest weight modules over the quantum group of the weight related to configuration [68,75]. Half-monodromy(=braiding) is given by the R-matrix (PR, where P is a permutation). Universal R-matrix is provided by Drinfeld's double [18].

Here is the organization of the paper. In sections $1-3$ we recall the multivalued form, the distinguished cycle Δ, and the normalization constant of ref. [43]. In section 4 we recall the version of quantum group used in refs. [68, 75] for the explicit version of Drinfeld-Kohno's theorem: half-monodromy=R-matrix [42]. In Section 5 we represent the distinguished cycle Δ as an element of the corresponding tensor product and check the monodromy properties. The cycle Δ has the meaning of q-antisymmetric tensor, see Theorem 5.7 below, and corresponds to a particular conformal block, Figure 10. Finally, we discuss the properties of the tensor product with vector representation of $sl(n + 1)$.

0.1. Notations

$\alpha_1, \alpha_2, \ldots, \alpha_n$ - simple roots of the root system of type A_n
$\quad \mathcal{R}_+$ - set of positive roots
$\quad \mathcal{R}$ root system of type A_n
$\quad \delta = \frac{1}{2} \sum_{\alpha \in \mathcal{R}_+} \alpha$ -halfsum of positive roots
$\quad k$- complex parameter: $k \in \mathbb{C}$

$$ \rho = \frac{k}{2} \sum_{\alpha \in \mathcal{R}_+} \alpha $$

$\varkappa = \frac{-1}{k}$
$q = exp(\frac{2\pi i}{\varkappa})$
$\omega_1, \omega_2, \ldots, \omega_n$ are fundamental weights: $(\omega_i, \alpha_j) = \delta_{ij}$
R- R-matrix
Δ - distinguished cycle integral over which provides the zonal spherical function of type A_n
$\quad \Delta$ - comultiplication in quantum group

1. Multivalued form and discriminantal configuration

Consider the following set of variables:
$\quad z_l, \quad l = 1, \ldots, n + 1, \ t_{ij}, \ i = 1, \ldots, j, \ j = 1, \ldots, n.$
$\quad$ Variables z_l have meaning of arguments, while variables t_{ij} are variables of integration.
$\quad$ It is convenient to organize variables z_l, t_{ij} in the form of a pattern, cf. Figure 1. The idea of such an organization is borrowed from Gelfand-Tsetlin patterns [28].

Definition 1.1 Consider the following multivalued form $w(z,t)$:

$$w(z,t) = \prod_{i=1}^{n+1} z_i^{\lambda_1 + \frac{kn}{2}} \prod_{i_1 > i_2} (z_{i_1} - z_{i_2})^{1-2k}$$

$$\times \prod_{l=1}^{n+1} \prod_{i=1}^{n} (z_l - t_{i,n})^{k-1}$$

$$\times \prod_{j=1}^{n-1} \prod_{i_1=1}^{j+1} \prod_{i=1}^{j} (t_{ij} - t_{i_1,j+1})^{k-1}$$

$$\times \prod_{j=2}^{n} \prod_{i_1 > i_2} (t_{i_1,j} - t_{i_2,j})^{2-2k}$$

$$\times \prod_{j=1}^{n} \prod_{i=1}^{j} t_{ij}^{\lambda_{n-j+2} - \lambda_{n-j+1} - k} \quad dt_{11} dt_{12} dt_{22} \dots dt_{nn} . \qquad (2)$$

Here k is a complex parameter, the 'half multiplicity' of a root, $\lambda_1, \dots, \lambda_{n+1}$ are complex parameters subject to the condition

$$\lambda_1 + \lambda_2 + \dots + \lambda_{n+1} = 0 .$$

The form $w(z,t)$ was considered in [44], where it was proved that integrals of the form $w(z,t)$ over appropriate cycles (and cycles were explicitly described) for generic λ, k provide the $(n+1)!$-dimensional space of solutions of the Sekiguchi-Debiard system of differential equations (Heckman-Opdam system of differential equations in the case of the root system of type A_n).

Before proceeding further we would like to make a convention.

Convention 1.2 A complex number z can be represented as $z = re^{i\alpha}$, where r, α are real numbers, $r \geq 0$. r is called the absolute value of z, while α is called the phase of z. When we say that the phase of a complex number z is equal to 0, we mean that $\alpha = 0$, or the number itself is real and nonnegative.

1.3 Configuration Let $m = \frac{n(n+1)}{2}$. Consider $(n+1+m)$-dimensional complex space $\mathbb{C}^{n+1+m}$ with coordinates $z_1, z_2, \dots, z_{n+1}, t_{11}, t_{12}, t_{22}, \dots, t_{nn}$. Let's delete the following hyperplanes:

$$t_{i_1,j} - t_{i_2,j} = 0 \quad i_1 < i_2, \; j = 1, \dots, n$$

$$t_{i_1,j} - t_{i_2,j+1} = 0 \quad j = 1, \dots, n-1$$

$$z_{i_1} - t_{i_2,n} = 0 \quad i_1 = 1, \dots, n+1; \; i_2 = 1, \dots, n$$

$$z_1 \qquad z_2 \qquad \cdots \qquad \cdots \qquad z_{n+1}$$

$$t_{1,n} \qquad t_{2,n} \qquad \cdots \qquad t_{n,n}$$

$$\cdots \qquad \cdots \qquad \cdots$$

$$t_{1,2} \qquad t_{2,2}$$

$$t_{1,1}$$

Figure 1: *Variables organized in a pattern.*

$$t_{ij} = 0 \quad i = 1,\dots,j; \; j = 1,\dots,n$$
$$z_i - z_j = 0 \quad i < j$$
$$z_i = 0 \quad i = 1,\dots,n+1.$$

Denote the complement of $\mathbb{C}^{m+n+1}$ to the union of above the hyperplanes by U_{n+1+m}.

Denote by *Loc* the trivial 1-dimensional bundle over U_{n+1+m} with the integrable connection ∇ with the connection form

$$\sum_{j,i}(k-1)\frac{d(t_{jn}-z_i)}{t_{jn}-z_i} + \sum_{i<j}(1-2k)\frac{d(z_i-z_j)}{z_i-z_j} + \sum_i (\lambda_1 + \frac{kn}{2})\frac{dz_i}{z_i}$$

$$+ \sum_{j,i_1<i_2}(2-2k)\frac{d(t_{i_1,j}-t_{i_2,j})}{t_{i_1,j}-t_{i_2,j}} + \sum_{j=1}^{n-1}\sum_{i_1,i_2}(k-1)\frac{d(t_{i_1,j}-t_{i_2,j+1})}{t_{i_1,j}-t_{i_2,j+1}}$$

$$+ \sum_{i,j}(\lambda_{n+2-j}-\lambda_{n+1-j}-k)\frac{dt_{ij}}{t_{ij}}.$$

Denote by S the local system of horizontal sections of ∇. Consider the projection on the first $n+1$ coordinates $\mathbb{C}^{n+1+m} \to \mathbb{C}^{n+1}$. For $z = (z_1, z_2, \dots, z_{n+1})$ such that $z_i \neq z_j$ for all $i < j$, set

$$U(z) = \left\{ (\tilde{z}, t) \in U_{n+1+m} | \tilde{z} = z \right\}.$$

Restrictions of *Loc*, S to $U(z)$ are denoted by $Loc(z), S(z)$. Denote by S^* the dual local system and consider the homology of $U(z)$ with coefficients in S^* (extended by ! as explained in ref. ([68]). Configuration is preserved under the action of the product of symmetric groups:

$$\Sigma = S_n \times S_{n-1} \times \dots \times S_2,$$

where S_j permutes $t_{1j}, t_{2j}, \ldots, t_{jj}$. Then following V. Schechtman and A. Varchenko [68] we consider the antiinvariant part of the homology group with respect to the action of Σ: $H_{*!,m}(U(z), S^*(z))^-$.

Strictly speaking, according to ref. [68] one should remove all the hyperplanes $z_i = t_{jl}$ and $t_{ij} = t_{i',j'}$. So consider the complement of $\mathbb{C}^{n+m+1}$ to the union of all the hyperplanes $z_i = 0$, $i = 1, \ldots, n+1$; $z_i = z_j$, $1 \leq i < j \leq n+1$, $t_{ij} = 0$, $i = 1, \ldots, j$, $j = 1, \ldots, n$; $t_{ij} = t_{i'j'}$, $i = 1, \ldots, j, j = 1, \ldots, n$, $i = 1, \ldots, j'$, $j' = 1, \ldots, n$; $z_i = t_{jl}$, $i = 1, \ldots, n+1$, $j = 1, \ldots, l$, $l = 1, \ldots, n$ and denote it by $\hat{U}_{n+m+1}$.

Set

$$\hat{U}(z) = \left\{ (\tilde{z}, t) \in \hat{U}_{n+m+1} | \tilde{z} = z \right\}$$

Consider now $H_{*!,m}(\hat{U}(z), S^*(z))^-$. We have a natural inclusion $\hat{U} \subset U$ which provides a natural morphism

$$\pi : H_{*!,m}(\hat{U}(z), S^*(z))^- \rightarrow H_{*!,m}(U(z), S^*(z))^- . \tag{3}$$

We will find an element of $H_{*!,m}(\hat{U}, S^*)^-$ (Theorem 5.7 parts 2 and 3 below) whose image in $H_{*!,m}(U, S^*)^-$ is the same as of cycle Δ.

For the geometric definition of the cycle Δ, see Section 2 below.

Remarkably, in order to calculate the cohomology group of the local system of the complement to a finite set of hyperplanes in the nonresonance case, one can use the finite-dimensional complex of hypergeometric forms in the spirit of Arnold-Orlik-Solomon, cf. [5,70].

2. The distinguished cycle Δ

Assume that $z_1, z_2, \ldots, z_{n+1}$ are real and

$$0 < z_1 < z_2 < \ldots < z_{n+1}.$$

Definition 2.1 Define a cycle $\Delta = \Delta(z)$ by the following inequalities:

$t_{i,j+1} \leq t_{ij} \leq t_{i+1,j+1}$ and
$z_i \leq t_{in} \leq z_{i+1}$ [43], Definition 2.1.

Define a form $\omega_\Delta(z,t)$ as

$$\omega_\Delta(z,t) = \prod_{i=1}^{n+1} z_i^{\lambda_1 + \frac{kn}{2}} \prod_{i_1 > i_2} (z_{i_1} - z_{i_2})^{1-2k}$$

$$\times \prod_{i \le l} (z_l - t_{i,n})^{k-1} \prod_{i > l} (t_{i,n} - z_l)^{k-1}$$

$$\times \prod_{j=1}^{n-1} \prod_{i_1 > i_2} (t_{i_1,j} - t_{i_2,j+1})^{k-1} \prod_{i_2 \ge i_1} (t_{i_2,j+1} - t_{i_1,j})^{k-1}$$

$$\times \prod_{j=2}^{n} \prod_{i_1 > i_2} (t_{i_1,j} - t_{i_2,j})^{2-2k}$$

$$\times \prod_{j=1}^{n} \prod_{i=1}^{j} t_{ij}^{\lambda_{n-j+2} - \lambda_{n-j+1} - k} \quad dt_{11} dt_{12} dt_{22} \ldots dt_{nn}$$

It is assumed that phases of all factors in the formula for ω_Δ are equal to zero if k and $\lambda_1, \lambda_2, \ldots, \lambda_{n+1}$ are real and then antisymmetrize with respect to Σ. In other words, we choose the section of the local system to be positive over Δ if λ, k are real. This geometric definition of cycle Δ is justified by Theorem 5.7 below, i.e., Δ really defines an element of $H_{*!,m}(U, S^*)^-$. This geometric definition is motivated by the classical calculations of I. Gelfand and M. Naimark of the zonal spherical function for $SL(n+1, \mathbb{C})$ [27]. The distinguished cycle Δ should be probably called the Gelfand-Naimark cycle.

If parameters $\lambda_j - \lambda_{j-1} - k$ are nonnegative integers for $j = 2, \ldots, n+1$, then the zonal spherical function as the hypergeometric series terminates and becomes the Jack polynomial.

In order to define geometrically an element of $H_{*!,m}(\hat{U}, S^*)^-$ whose image in $H_{*!,m}(U, S^*)^-$ is cycle Δ we should bend Δ a little bit into the complex domain as it is shown in Figure 2, and then antisymmetrize with respect to Σ. This cycle will be denoted by $\tilde{\Delta}$.

3. Analytic considerations

Let

$$\lambda_{n-j+2} - \lambda_{n-j+1} - k = 0$$

for $j = 1, \ldots, n$ and $\lambda_1 + \frac{kn}{2} = 0$, i.e., we kill an affine part of the integral. Then in these hypotheses

$$\int_{\Delta(z)} \omega_\Delta(z,t) = \frac{\Gamma(k)\Gamma(k)^2 \ldots \Gamma(k)^{n+1}}{\Gamma(k)\Gamma(2k) \ldots \Gamma((n+1)k)},$$

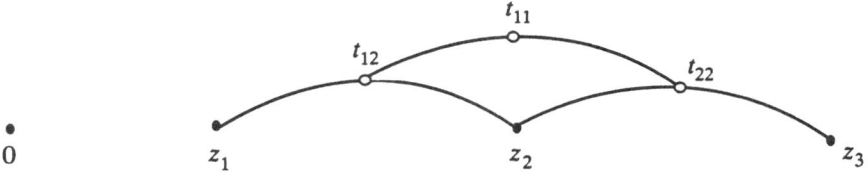

Figure 2: *Cycle Δ bent into the complex domain (in the case of the root system of type A_2).*

see ref. [43] Theorem 1.5 and Remark 1.6.
Following the classical work of [27] let

$$\tau_{ij} = \frac{\displaystyle\prod_{i_1=1}^{j-1} (t_{i_1,j-1} - t_{ij})}{\displaystyle\prod_{i_1 \neq i} (t_{i_1,j} - t_{ij})} \quad ,$$

$i = 1, \ldots, j$; $j = 2, \ldots, n$. Note that $\displaystyle\sum_{i=1}^{j} \tau_{ij} = 1$, and

$$\frac{D(\tau_{1,j}, \ldots, \tau_{j-1,j})}{D(t_{1,j-1}, \ldots, t_{j-1,j-1})} = \frac{\displaystyle\prod_{1 \leq i < k \leq j-1} (t_{i,j-1} - t_{k,j-1})}{\displaystyle\prod_{1 \leq i < p \leq j} (t_{ij} - t_{pj})}$$

[see [27] for the details]. Let also

$$\tau_{i,n+1} = \frac{\displaystyle\prod_{i=1}^{n} (t_{i_1,n} - z_i)}{\displaystyle\prod_{i_1 \neq i} (z_{i_1} - z_i)} \qquad i = 1, \ldots, n+1.$$

One has $\displaystyle\sum_{i=1}^{n+1} \tau_{i,n+1} = 1$ and

$$\frac{D(\tau_{1,n+1}, \ldots, \tau_{n,n+1})}{D(t_{1,n}, \ldots, t_{n,n})} = \frac{\displaystyle\prod_{1 \leq i < k \leq n} (t_{i,n-1} - t_{k,n})}{\displaystyle\prod_{1 \leq i < p \leq n+1} (z_i - z_p)} .$$

In variables τ_{ij} $\quad i = 1, \ldots, j-1, \quad j = 2, \ldots, n+1$ integral $\int_\Delta \omega_\Delta$ is written as

$$\int_\Delta \omega_\Delta = \int \prod_{j=1}^{n+1} ((\tau_{1j}\tau_{2j}\ldots\tau_{j-1,j})(1 - \tau_{1j} - \ldots - \tau_{j-1,j}))^{k-1}$$
$$\times \, d\tau_{12}d\tau_{13}d\tau_{23}\ldots d\tau_{1,n+1}\ldots d\tau_{n,n+1} \ .$$

Remarkably in variables τ_{ij} the integration is being performed over one-dimensional simplex times two-dimensional simplex times and so on, times n-dimensional simplex. Here the one-dimensional simplex corresponds to a line in the two-dimensional plane, the two-dimensional simplex corresponds to the two-dimensional plane in three-dimensional plane, and so on. Thus, in some sense the integral remembers the flag manifold.

So using Dirichlet's formula, one gets

$$\int_\Delta \omega_\Delta = \frac{\Gamma(k)\Gamma(k)^2\ldots\Gamma(k)^{n+1}}{\Gamma(k)\Gamma(2k)\ldots\Gamma((n+1)k)} \ .$$

The constant does not depend on z_i at all and surely remains the same under analytic continuation. This is nontrivial since form $\omega = \omega(z,t)$ and cycle $\Delta = \Delta(z)$ do depend on $z = (z_1, z_2, \ldots, z_{n+1})$.

Remark 3.1 In view of sections 4 and 5 below, killing off the affine part (Mellin part) in the integral corresponds to erasing the first factor in the tensor product of irreducible highest weight modules over quantum group.

4. The quantum group $U_q(sl(n+2))$

Quantum groups were introduced by Drinfeld [18], Jimbo [41], Kulish, Reshetikhin, Sklyanin [49]. We briefly recall the necessary material from [68,75] and refer directly to the above references for more details.

4.1 Root system

Let $\mathbb{R}^{n+2}$ be Euclidean $(n+2)$-dimensional vector space with inner product $(.,.)$ and with $g_0, g_1, \ldots, g_{n+1}$ as the orthonormal basis. Let's realize simple roots of root system of type A_n as $\alpha_i = g_i - g_{i+1}$ for $i = 1, \ldots, n$. Set also $\alpha_0 = g_0 - g_1$.

In particular, one has

$$(\alpha_i, \alpha_i) = 2$$
$$(\alpha_i, \alpha_j) = 0 \quad \text{for} \quad |i - j| > 1$$
$$(\alpha_i, \alpha_j) = -1 \quad \text{for} \quad |i - j| = 1$$

Set also

$$\alpha^\vee = \frac{2\alpha}{(\alpha, \alpha)}.$$

4.2. The quantum group $U_q(sl(n+2))$

Consider $\mathbb{C}$-algebra with generators $e_i, f_i,$ and $K_i^{\frac{1}{2}}, K_i^{\frac{-1}{2}}, i = 0, \ldots, n,$ subject to the relations:

$$K_j^{\frac{1}{2}} e_i = q^{\frac{(\alpha_i, \alpha_j^\vee)}{4}} e_i K_j^{\frac{1}{2}}$$

$$K_j^{\frac{1}{2}} f_i = q^{-\frac{(\alpha_i, \alpha_j^\vee)}{4}} f_i K_j^{\frac{1}{2}}$$

$$[e_i, f_j] = (K_i - K_i^{-1})\delta_{ij}$$

$$K_i^{\pm\frac{1}{2}} K_j^{\frac{1}{2}} = K_j^{\frac{1}{2}} K_i^{\pm\frac{1}{2}}$$

$$K_i^{\frac{1}{2}} K_i^{\frac{-1}{2}} = K_i^{\frac{-1}{2}} K_i^{\frac{1}{2}} = 1$$

Comultiplication is defined by

$$\Delta(K_i^{\pm\frac{1}{2}}) = K_i^{\pm\frac{1}{2}} \otimes K_i^{\pm\frac{1}{2}}$$

$$\Delta(f_i) = f_i \otimes K_i^{\frac{1}{2}} + K_i^{-\frac{1}{2}} \otimes f_i$$

$$\Delta(e_i) = e_i \otimes K_i^{\frac{1}{2}} + K_i^{-\frac{1}{2}} \otimes e_i$$

The following are the quantum Serre relations:

$$f_i^2 f_{i+1} - (q^{\frac{1}{2}} + q^{-\frac{1}{2}})f_i f_{i+1} f_i + f_{i+1} f_i^2 = 0$$

$$f_{i+1}^2 f_i - (q^{\frac{1}{2}} + q^{-\frac{1}{2}})f_{i+1} f_i f_{i+1} + f_i f_{i+1}^2 = 0$$

$$f_i f_j = f_j f_i \quad \text{for} \quad |i - j| \neq 1$$

$$e_i^2 e_{i+1} - (q^{\frac{1}{2}} + q^{-\frac{1}{2}})e_i e_{i+1} e_i + e_{i+1} e_i^2 = 0$$

$$e_{i+1}^2 e_i - (q^{\frac{1}{2}} + q^{-\frac{1}{2}})e_{i+1} e_i e_{i+1} + e_i e_{i+1}^2 = 0$$

$$e_i e_j = e_j e_i \quad \text{for} \quad |i - j| \neq 1.$$

The $\mathbb{C}$-algebra generated by elements $e_i, f_i, K_i^{\frac{1}{2}}, K_i^{-\frac{1}{2}}, i = 0, \ldots, n$, subject to the above relations and with the comultiplication will be referred to as the quantum group $U_q(sl(n+2))$. Antipode S and counit ϵ are defined appropriately [18,41,68].

4.3. Verma module

For $\Lambda \in span\{\alpha_i\}$ denote by $M(\Lambda)$ the Verma module generated over the quantum group without Serre's relations by a single vector v subject to relations $K_i^{\frac{1}{2}} v = q^{\frac{(\Lambda, \alpha_i^\vee)}{4}} v$, $e_i v = 0$ for all i.

$$M(\Lambda)_\mu = \left\{ x \in M(\Lambda) \,|\, K_i^{\frac{1}{2}} x = q^{\frac{(\Lambda - \mu, \alpha_i^\vee)}{4}} x \right\}.$$

Set $\tau(e_i) = f_i$, $\tau(f_i) = e_i$, $\tau(K_i^{\pm\frac{1}{2}}) = K_i^{\pm\frac{1}{2}}$ on generators and extend τ as the algebra antihomomorphism. Put $M(\Lambda)^* = \oplus_\mu M(\Lambda)_\mu^*$. Define the structure of quantum group module on $M(\Lambda)^*$ by the rule $(g\phi, x) = (\phi, \tau(g)x)$.

In the Verma module without Serre's relations, all monomials $f_{i_1} \cdots f_{i_l} v$ are linearly independent and form a basis. By $(f_{i_1} \cdots f_{i_l} v)^*$ we denote elements of the corresponding dual basis.

4.4. Contravariant form

Contravariant form (Shapovalov form) [68] on a Verma module with highest weight vector v is defined such that

$$S(v, v) = 1$$

and

$$S(f_i x, y) = S(x, e_i y)$$

for all i, x, y.

The contravariant form S defines the homomorphism of modules:

$$S : M(\Lambda) \to M(\Lambda)^*.$$

Example 1.

$$S : v \mapsto v^*.$$

Example 2.

$$S : f_1 v \mapsto \left(q^{\frac{(\Lambda, a_1)}{2}} - q^{-\frac{(\Lambda, a_1)}{2}} \right) (f_1 v)^*,$$

see also Figure 3.

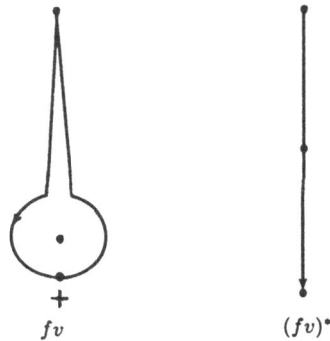

$$fv \qquad\qquad (fv)^*$$

Figure 3: *Chains and quantum group [68,75],[34]. Phase is chosen to be zero at the point marked with +.*[1]

Let

$$L(\Lambda) = M(\Lambda)/\mathrm{Ker}S$$

be irreducible module over the quantum group with Serre's relations with highest weight Λ.

4.5 R-matrix

R-matrix is defined by the following expression:

$$R = \sum_\mu q^{\frac{\Omega_0}{2} + \frac{1}{4}(\mu \otimes 1 - 1 \otimes \mu) + d(\mu)} \Omega_\mu$$

see [18]. Here Ω_0 is the element corresponding to the inner product $(.,.)$; for

$$\mu = l_0 \alpha_0 + l_1 \alpha_1 + \ldots + l_n \alpha_n,$$

where $l_0, l_1, \ldots, l_n$ are nonnegative integers, $d(\mu) \in \mathbb{C}$ is a constant defined as follows: represent $\mu = \sum l_i \alpha_i$ as a sum of simple roots with repetitions $\mu = \alpha_{i_1} + \alpha_{i_2} + \ldots + \alpha_{i_n}$. Then

$$d(\mu) = - \sum_{p \leq q} \frac{(\alpha_{i_p}, \alpha_{i_q})}{4}.$$

---

[1]The choice of comultiplication is dictated by the choice that the phase is equal to zero at the point marked with +.

Ω_μ is a canonical element [18].

R defines a linear operator

$$R : M \otimes M' \to M \otimes M'.$$

The following diagram is commutative :

$$
\begin{array}{ccc}
M \otimes M' & \xrightarrow{\;R\;} & M \otimes M' \\
{\scriptstyle s}\downarrow & & \downarrow{\scriptstyle s} \\
M^* \otimes M'^* & \xrightarrow[\;R^*\;]{} & M^* \otimes M'^*
\end{array}
$$

see Theorem 7.6.8 of ref. [75]. R induces a homomorphism of irreducible highest weight modules:

$$R : L(\Lambda(1)) \otimes L(\Lambda(2)) \to L(\Lambda(1)) \otimes L(\Lambda(2))$$

which will be also denoted by R.

Denote by P the transposition of two factors in the tensor product

$$P : M \otimes M' \to M' \otimes M.$$

5. Quantum group and cycle Δ

5.1 Data

We are going to make a homological analysis of the distinguished (Gelfand-Naimark) cycle Δ and to check its monodromy properties using the quantum group argument [68, 75].

Take now a different indexation of variables t_{ij}. Namely, we are going to use $\{t_i{}^{(j)} \mid j = 1,\dots,n; \; i = j,\dots,n\}$. Set also $z_0 = 0$ (affine configuration). Consider the following multivalued form:

$$
\begin{aligned}
\Omega(z,t) &= \prod (z_0 - z_i)^{\frac{(\Lambda(0),\Lambda(i))}{\varkappa}} \prod (z_i - z_j)^{\frac{(\Lambda(i),\Lambda(j))}{\varkappa}} \\
&\times \prod (z_i - t_l{}^{(j)})^{\frac{(\Lambda(i),-\alpha_j)}{\varkappa}} \prod (z_0 - t_l{}^{(j)})^{\frac{(\Lambda(0),-\alpha_j)}{\varkappa}} \\
&\quad \prod (t_l{}^{(j)} - t_{l'}{}^{(j')})^{\frac{(-\alpha_j,-\alpha_{j'})}{\varkappa}} dt_1{}^{(1)} \dots dt_n{}^{(n)}
\end{aligned}
\tag{3}
$$

Integrals of forms of this type are considered in refs. [68,75]. Now we would like to specialize $\Lambda(0), \Lambda(1), \dots, \Lambda(n+1), \alpha_1, \alpha_2, \dots, \alpha_n$ as follows.

Recall that $\mathbb{R}^{n+2}$ is an $(n+2)$-dimensional Euclidean vector space with $g_0, g_1, \ldots, g_{n+1}$ as the orthonormal basis. For $i = 1, \ldots, n+1$ set

$$\Lambda(i) = \Lambda = g_1 - g_0$$

e., to each variable $z_i, i = 1, \ldots, n+1$ assign the same vector $g_1 - g_0$.

Recall that simple roots of the root system of type A_n are realized as follows:

$$\alpha_i = g_i - g_{i+1}, \quad \text{for} \quad i = 1, \ldots, n.$$

Remark 5.2   Note: the projection of $\Lambda(i)$ on the span of $\alpha_i, i = 1, \ldots, n$ is exactly the first fundamental weight ω_1, i.e., $(\Lambda(i), \alpha_j) = \delta_{1j}$, where δ_{1j} is a Kronecker's delta, but $(\Lambda(i), \Lambda(i)) = 2$. This is not very important since it changes only the power of $\prod(z_i - z_j)$ before the integral, but does not change the fiber $H_{*!,m}(U, S^*)^-$.

The form $\omega(z, t)$ can be also rewritten as

$$\omega(z, t) = \prod(z_i - z_j)^{1-2k} \prod z_i^{(\lambda + \rho, \omega_1)}$$
$$\times \prod t_i^{(j)(\lambda + \rho, -\alpha_j)} \prod (z_l - t_i^{(j)})^{(\omega_1, -\alpha_j)(1-k)}$$
$$\times \prod (t_i^{(j)} - t_{i'}^{(j')})^{(-\alpha_j, -\alpha_{j'})(1-k)} dt_1^{(1)} \ldots dt_n^{(n)} \tag{4}$$

vector $\rho = \rho(k)$ is defined immediately below. The integral representation (4) allows us to prove that a basis of solutions to the Sekiguchi-Debiard system of differential equations is provided by certain conformal blocks of the WA_n-algebra, see [46].

Let $\mathcal{R}$ denotes the root sytem of the type A_n with simple roots $\alpha_1, \alpha_2, \ldots, \alpha_n$ as before. $\mathcal{R}_+$ denotes the set of positive roots. Let δ be half the sum of positive roots:

$$\delta = \frac{1}{2} \sum_{\alpha \in \mathcal{R}_+} \alpha$$

$$\rho = \rho(k) = \frac{k}{2} \sum_{\alpha \in \mathcal{R}_+} \alpha$$

Set also $\varkappa = -\frac{1}{k}$. Let λ belong to the span of $\alpha_1, \ldots, \alpha_n$. Set $\Lambda(0) = \varkappa\lambda - \delta$, so that

$$\frac{\Lambda(0)}{\varkappa} = \lambda + \rho.$$

The multivalued form $\omega(z, t)$ of section 1 differs from $\Omega(z, t)$ in the above setting only by some meromorphic factor which does not contribute to the monodromy and thus can be omitted for the purposes of this section.

Set $q = \exp(\frac{2\pi i}{\varkappa})$. We assume that parameter $\varkappa$ is irrational.

Remark 5.3 Note: our form is somewhat different from the A. Matsuo's form of ref. [56], in particular, in the setting of [56] the cycle Δ will not serve as a cycle for integration for zonal spherical function (also we do not have the "complicated meromorphic factor").

5.4. The most trivial example

Before proceeding further we want to consider the most trivial example. Namely, consider

$$(z_1 - z_2)^{\frac{(\Lambda(1),\Lambda(2))}{\varkappa}}$$

If z_2 goes around z_1 counterclockwise, then this function gains the factor $\exp(\frac{2\pi i}{\varkappa}(\Lambda(1),\Lambda(2)))$. If z_2 goes halfway around z_1, then the function gains the factor $\exp(\frac{\pi i}{\varkappa}(\Lambda(1),\Lambda(2))) = q^{\frac{(\Lambda(1),\Lambda(2))}{2}}$. At the same time consider the tensor product $v_1 \otimes v_2$ of highest weight vectors of modules of the corresponding quantum group of weights $\Lambda(1)$ and $\Lambda(2)$, respectively. Let R be the R-matrix; then

$$R(v_1 \otimes v_2) = q^{\frac{\Omega_0}{2}} v_1 \otimes v_2,$$

where Ω_0 is the canonical element corresponding to inner product, i.e.,

$$q^{\frac{\Omega_0}{2}} v_1 \otimes v_2 = q^{\frac{(\Lambda(1),\Lambda(2))}{2}} v_1 \otimes v_2$$

in agreement with the above considerations.

5.5. Case of the root system of type A_1 $(n = 1)$

Consider the tensor product of three dual Verma modules over $U_q(sl(3))$ (simple roots α_0, α_1)

$$M(\Lambda(0))^* \otimes M(\Lambda(1))^* \otimes M(\Lambda(2))^*.$$

Let $v_0 \otimes v_1 \otimes v_2$ be the tensor product of highest weight vectors. Then cycle Δ is encoded as

$$v_\Delta = -q^{\frac{(\Lambda(1),-\alpha_1)}{4}} v_0^* \otimes (f_1 v_1)^* \otimes v_2^* + q^{-\frac{(\Lambda(2),-\alpha_1)}{4}} v_0^* \otimes v_1^* \otimes (f_1 v_2)^* =$$
$$- q^{-\frac{1}{4}} v_0^* \otimes (f_1 v_1)^* \otimes v_2^* + q^{\frac{1}{4}} v_0^* \otimes v_1^* \otimes (f_1 v_2)^*. \quad (5)$$

Here $*$ is as in Section 4.3.

Consider the action of R-matrix on the second and third component of the tensor product

$$R: M(\Lambda(1))^* \otimes M(\Lambda(2))^* \to M(\Lambda(1))^* \otimes M(\Lambda(2))^*.$$

Recall that P denotes the permutation of factors:

$$P: M(\Lambda(1))^* \otimes M(\Lambda(2))^* \to M(\Lambda(2))^* \otimes M(\Lambda(1))^*.$$

Now one can utilize formulas of example 1 of [68]. Then

$$PR: v_\Delta \mapsto (-1)v_\Delta$$

And so

$$(PR)^2 : v_\Delta \mapsto v_\Delta.$$

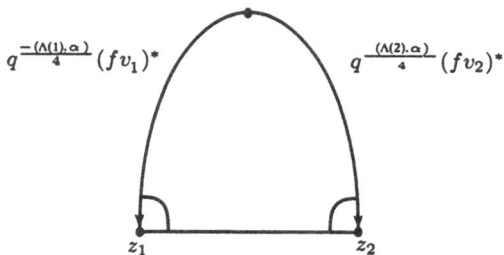

Figure 4: *Decomposition with the help of the quantum group cf. [68,75,34].*

5.6. Case of the root system of type A_n

Theorem 5.7 *Let $v_0, v_1, \dots, v_{n+1}$ be the highest weight vectors of Verma modules over the quantum group $U_q(sl(n+2))$ without Serre's relations with highest weights $\Lambda(0), \Lambda(1) = \Lambda(2) = \dots = \Lambda(n+1) = \Lambda$, respectively. Here $\Lambda(0), \Lambda$ are as in Section 5.1.*

1. Then the image of cycle $\tilde{\Delta}$ in $H_{*!, \frac{n(n+1)}{2}}(\hat{U}, S^*)^-$ is represented as an element of the tensor product of dual Verma modules $M(\Lambda(0))^* \otimes M(\Lambda(1))^* \otimes \dots \otimes M(\Lambda(n+1))^*$ as

$$v_\Delta = \sum_{w \in S_{n+1}} (-1)^{l(w)} q^{\frac{1}{4}[\frac{n(n+1)}{2} - 2l(w)]} v_0^* \otimes (f_{w(1)-1} \cdots f_2 f_1 v_1)^* \otimes \cdots$$

$$\otimes (f_{w(i)-1} \cdots f_2 f_1 v_i)^* \otimes \cdots$$

$$\otimes (f_{w(n+1)-1} \cdots f_2 f_1 v_{n+1})^*.$$

2. Let $\tilde{v}_\Delta$ be the following element of the tensor product of irreducible highest weight modules over the quantum group $U_q(sl(n+2))$ with Serre's relations $L(\Lambda(0)) \otimes L(\Lambda(1)) \otimes \cdots \otimes L(\Lambda(n+1))$:

$$\tilde{v}_\Delta = \sum_{w \in S_{n+1}} \frac{(-1)^{l(w)} q^{\frac{1}{4}[\frac{n(n+1)}{2} - 2l(w)]}}{(q^{\frac{1}{2}} - q^{-\frac{1}{2}})^{\frac{n(n+1)}{2}}}$$

$$\times v_0 \otimes f_{w(1)-1} \cdots f_2 f_1 v_1 \otimes \cdots$$

$$\otimes f_{w(i)-1} \cdots f_2 f_1 v_i \otimes \cdots$$

$$\otimes f_{w(n+1)-1} \cdots f_2 f_1 v_{n+1}$$

Then $S(\tilde{v}_\Delta) = v_\Delta$, where S is the contravariant (Shapovalov) form.

In particular, $e_i \tilde{v}_\Delta = 0$ for $i = 0, 1, \ldots n$. This implies that the image of the cycle $\tilde{\Delta}$ defines correctly an element of $H_{*!, \frac{n(n+1)}{2}}(\hat{U}, S^*)^-$ and thus the cycle Δ defines correctly an element of $H_{*!, \frac{n(n+1)}{2}}(U, S^*)^-$.

3. Let $\pi : H_{*!, \frac{n(n+1)}{2}}(\hat{U}, S^*)^- \to H_{*!, \frac{n(n+1)}{2}}(U, S^*)^-$ be a natural morphism corresponding to imbedding $\hat{U} \to U$. Then

$$\pi(\tilde{\Delta}) = \Delta.$$

Proof. The notion of a diagram is recalled in the next section. To prove the theorem we repeatedly use Figures 3 and 4. Also for the above theorem the Theorem 2.6 and Remark 2.7 of [44] are helpful: the number of arrows which are 'to the left' is equal to the length of the corresponding element of the Weyl group (also reproduced below in Theorem 5.3). In fact, using elementary decomposition of Figure 4 we get that each arrow to the right brings the factor $q^{\frac{1}{4}}$, while each arrow to the left brings the factor $(-1)q^{\frac{-1}{4}}$. The number of variables $t_i^{(j)}$ is equal to $\frac{n(n+1)}{2}$. The key point is that all the 'wrong' diagrams (cf. Figure 6) cancel each other because of the phase argument, cf. Figure 7. Here by the 'wrong' diagrams we mean the diagrams in which there are two arrows with the same target.

Vice-versa, this theorem might be considered as the quantum group explanation of Theorem 2.6 of [44] (number of arrows which are to the left $= l(w)$). $\qquad\Box$

5.8. Diagrams

The notion of a diagram was introduced in [15] in the context of the trigonometric Knizhnik-Zamolodchikov equations, and later a similar notion was introduced in [17] in the context of multidimensional determinants and discriminants. We will use the V. Dolotin's notion of a diagram

[17], in particular, we borrow the very convenient graphical notation (see Figures 1, 2, and 3) of [17].

Fix some positive integer n. Consider the set of $\frac{(n+1)(n+2)}{2}$ points, indexed by pairs of integers $\{(i,j)|\ i = 1,\ldots,j,\ j = 1,\ldots,n+1\}$. It is helpful to organize the points in the form of a pattern, so that points are divided in n rows, jth row is formed by points $\{(i,j)|\ i = 1,\ldots,j\}$; the point (i,j) is located under and between points $(i,j+1)$ and $(i+1,j+1)$ (Figure 5a).

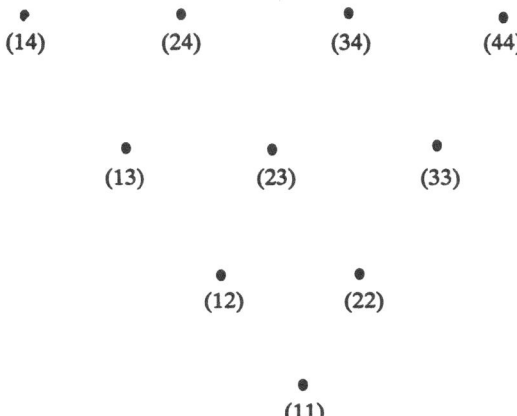

Figure 5A: $n = 3$. *Points (i,j) organized in a pattern; j is the number of the row, i is the number in the row.*

Now mark with a cross one point in each row. Let $\{(i_j,j)\}$ be the subset of marked points (Figure 5b).

Finally, draw an arrow for each point (i,j), $i = 1,\ldots,j,\ j = 1,\ldots,n$ with the source in this point (i,j) and target $tar(i,j)$ in the next $j+1$th row defined as

$$tar(i,j) = \begin{cases} (i,j+1), & \text{if } i < i_{j+1} \\ (i+1,j+1), & \text{if } i \geq i_{j+1}.\end{cases}^2$$

If $tar(i,j) = (i,j+1)$, then the arrow is called to the **left**; if $tar(i,j) = (i+1,j+1)$, then the arrow is called to the **right**. Note: neither arrow has a marked point as its target. In this way one obtains Figure 5c.

---

²Recall that $\{(i_j,\ j)\}$ is the set of marked points and $(i_{j+1},\ j+1)$ is the only marked point in $j+1$th row.

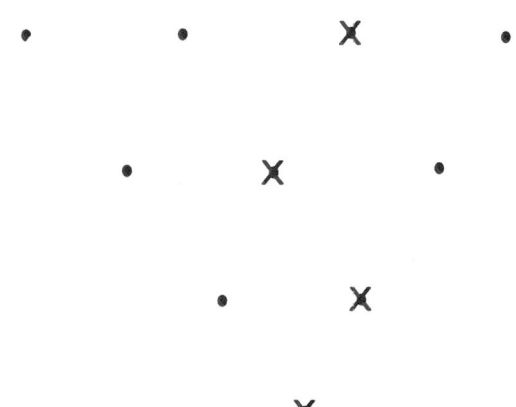

Figure 5B: *One point is marked in each row.*

Definition 5.9 A triple

$$(\{(i,\ j)\},\quad \{(i_j,\ j)\},\quad tar)$$

consisting of:

    set of points $\{(i,\ j)|\quad i=1,\dots,j\ ;\ j=1,\dots,n+1\}$,

    set of marked points $\{(i_j,\ j)\ |\quad j=1,\dots,n+1\}$

    and the function tar defined above will be called **a diagram.**

Remark 5.10 One can see that a diagram is determined by the set of marked points.

Definition 5.11 We describe the correspondence between diagrams and elements of the symmetric group as follows. Consider a diagram as an oriented graph and forget orientation. For $i=1,\dots,n+1$, define $w(i)$ as the number of points in the connected component of the point $(i, n+1)$.

    The symmetric group S_{n+1} has standard generators $\sigma_1,\ \sigma_2,\dots,\sigma_n$, where σ_i permutes i and $i+1$.

Definition 5.12 The length $l(w)$ of an element $w \in S_{n+1}$ is the minimal integer $p \geq 0$, such that w admits a presentation

$$w = \sigma_{i_1}\sigma_{i_2}\dots\sigma_{i_p}\ .$$

Any presentation of w as a product of $p = l(w)$ generators is called a reduced presentation.

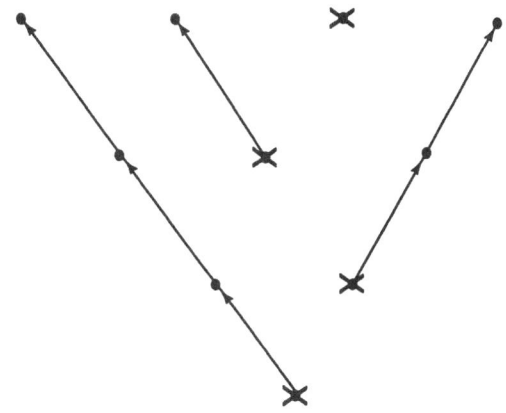

Figure 5C: *Example of a diagram.*

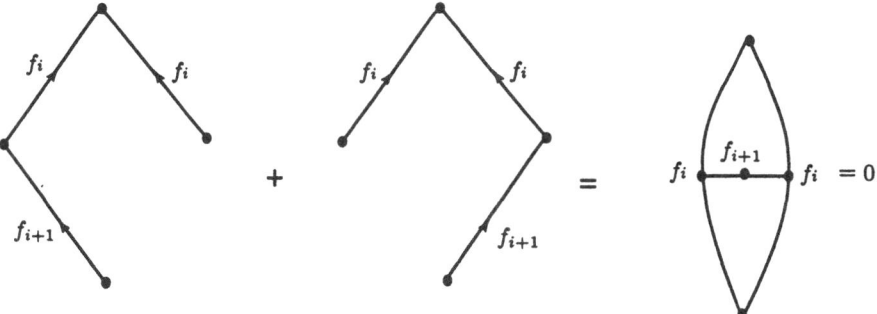

Figure 6: 'Wrong' diagrams cancel each other.

Theorem 5.13 [17] *Let a diagram*

$$(\{(i,j)| \quad i = 1,\dots,j, \; j = 1,\dots,n+1\}, \quad \{(i_j,j)| \quad j = 1,\dots,n+1\}, \; tar)$$

correspond to an element $w \in S_{n+1}$. Then the length $l(w)$ of an element w is equal to

$$l(w) = \sum_{j=1}^{n+1} (i_j - 1)$$

In other words, $l(w)$ is equal to the number of arrows which are to the left in the diagram corresponding to the element w, see Theorem 2.6 of [44].

Theorem 5.14 Let $\Lambda(1) = \Lambda(2) = \Lambda$, as in Theorem 5.7. Also, let v_1, v_2 be the highest weight vectors of Verma modules $M(\Lambda(1)), M(\Lambda(2))$

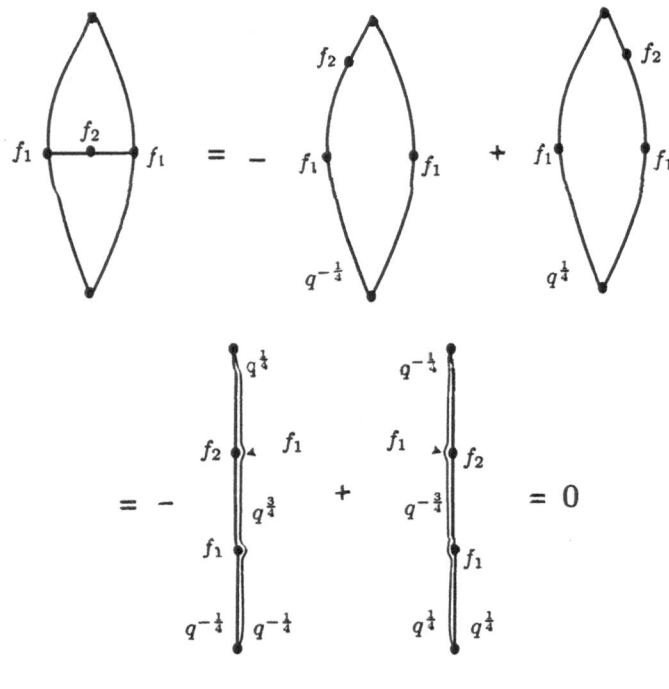

$$-q^{-\frac{1}{4}}(q^{\frac{1}{4}}f_1f_2f_1 + q^{\frac{3}{4}}f_2f_1f_1 + q^{-\frac{3}{4}}f_2f_1f_1)$$
$$+ q^{\frac{1}{4}}(q^{-\frac{1}{4}}f_1f_2f_1 + q^{-\frac{3}{4}}f_2f_1f_1 + q^{\frac{3}{4}}f_2f_1f_1) = 0$$

Figure 7: *Phase argument: decomposition into ordered chains — the total sum is an identical zero chain.*

over the quantum group $U_q(sl(n+2))$ without Serre's relations. Let $M(\Lambda(1))^*, M(\Lambda(2))^*$ be dual Verma modules. Let PR be the braiding:

$$PR : M(\Lambda(1))^* \otimes M(\Lambda(2))^* \to M(\Lambda(2))^* \otimes M(\Lambda(1))^*.$$

Then for $i > j$ one has

$$PR\left((f_if_{i-1}\ldots f_1v_1)^* \otimes (f_jf_{j-1}\ldots f_1v_2)^*\right) =$$
$$q^{\frac{1}{2}}(f_jf_{j-1}\ldots f_1v_2)^* \otimes (f_if_{i-1}\ldots f_1v_1)^* +$$
$$q^{\frac{1}{2}}(q^{\frac{1}{2}} - q^{-\frac{1}{2}})(f_if_{i-1}\ldots f_1v_2)^* \otimes (f_jf_{j-1}\ldots f_1v_1)^* \quad (6)$$

For $i < j$ one has

$$PR\left((f_if_{i-1}\ldots f_1v_1)^* \otimes (f_jf_{j-1}\ldots f_1v_2)^*\right) =$$
$$q^{\frac{1}{2}}(f_jf_{j-1}\ldots f_1v_2)^* \otimes (f_if_{i-1}\ldots f_1v_1)^*. \quad (7)$$

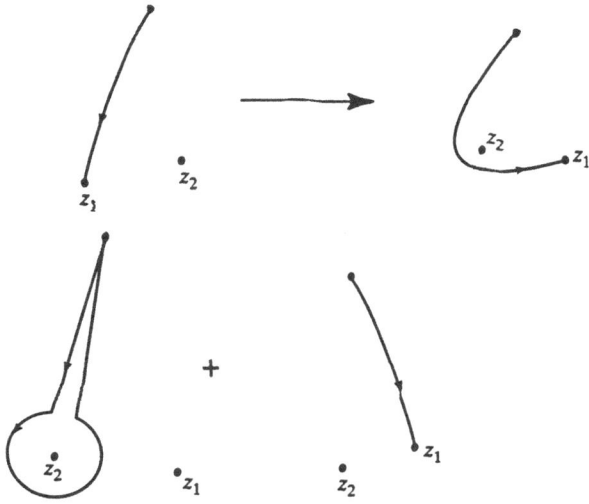

Figure 8: Result of the braiding.

Remark The monodronomy of the Heckman-Opdam hypergeometric functions was calculated by G. Heckman and E. Opdam [37]. In the case of the root system of type A_n, we can reobtain the monodronomy using our integral representation $\omega(z,t)$ for the solutions and the quantum group argument.

Proof. Theorem 5.14 is easily proved by contour manipulations, see Figure 7 as elementary, but for a typical example, see also Section 5.16 below.

$\square$

Corollary 5.15 Suppose $z_1(t), z_2(t), \ldots, z_{n+1}(t)$, $t \in [0,1]$ are closed loops on a complex plane, i.e., $z_1(0) = z_1(1), z_2(0) = z_2(1), \ldots, z_{n+1}(0) = z_{n+1}(1)$, such that $z_i(t) \neq z_j(t)$ for $i \neq j$. Let also $Re(z_i(t)) > 0$ for each $i = 1, \ldots, n+1$. Then the image of the homological class of the cycle $\tilde{\Delta}$ in $H_{*!, \frac{n(n+1)}{2}}(\hat{U}, S^*)^-$ is preserved under the monodromy along paths $z_i(t)$ and thus the homological class of Δ is preserved in $H_{*!, \frac{n(n+1)}{2}}(U, S^*)^-$ under such monodronomies.

Proof. In fact such monodromy can be produced as the composition of an even number of elementary braidings as in Theorem 5.14, each of them giving the factor -1. Braiding of z_1 with $z_0 = 0$ is forbidden by hypotheses.

$\square$

5.16. Using R-matrix for tensor product of two vector representations

The R-matrix for the tensor product of two vector representations of $sl(n+1)$ reads as

$$R = q^{-1/2(n+1)}\left\{ \sum_{i\neq j} E_{ii}\otimes E_{jj} + q^{\frac{1}{2}}\sum_{i} E_{ii}\otimes E_{ii} + (q^{\frac{1}{2}}-q^{-\frac{1}{2}})\sum_{i<j} E_{ij}\otimes E_{ji}\right\},$$

see [18, 41]. The vector representation of $sl(n+1)$ has a natural basis $e_1, e_2, \ldots, e_{n+1}$ with highest weight vector e_1, and here E_{ij} are the matrix units

$$E_{ij}e_k = \delta_{kj}e_i,$$

where δ_{kj} is Kronecker's delta. In our case (setting of Theorem 5.14) the interesting part of the R-matrix can be easily obtained from the above by multiplying by $q^{\frac{n+2}{2(n+1)}}$:

$$\hat{R} = q^{\frac{1}{2}}\left\{ \sum_{i\neq j} E_{ii}\otimes E_{jj} + q^{\frac{1}{2}}\sum_{i} E_{ii}\otimes E_{ii} + (q^{\frac{1}{2}}-q^{-\frac{1}{2}})\sum_{i<j} E_{ij}\otimes E_{ji}\right\}.$$

Then one immediately verifies that

$$\hat{R}(e_k\otimes e_l) = q^{\frac{1}{2}}e_k\otimes e_l \quad \text{if} \quad k<l$$

and

$$\hat{R}(e_k\otimes e_l) = q^{\frac{1}{2}}e_k\otimes e_l + q^{\frac{1}{2}}(q^{\frac{1}{2}}-q^{-\frac{1}{2}})e_l\otimes e_k \quad if \quad l<k$$

Also,

$$\hat{R}(e_k\otimes e_k) = qe_k\otimes e_k.$$

Let P be the transposition, i.e.,

$$Pe_i\otimes e_j = e_j\otimes e_i$$

Then one immediately obtains that for $k<l$,

$$P\hat{R}(q^{\frac{1}{4}}e_k\otimes e_l - q^{-\frac{1}{4}}e_l\otimes e_k) = (-1)(q^{\frac{1}{4}}e_k\otimes e_l - q^{-\frac{1}{4}}e_l\otimes e_k)$$

i.e.,

$$q^{\frac{1}{4}}e_k\otimes e_l - q^{-\frac{1}{4}}e_l\otimes e_k$$

is an eigenvector of $P\hat{R}$ with eigenvalue -1 !   □

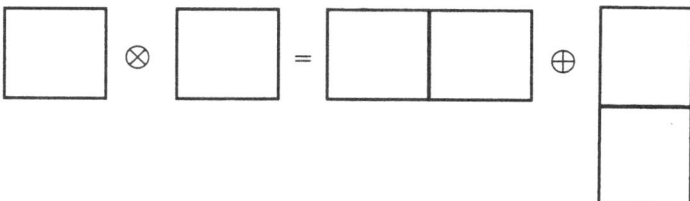

Figure 9: *Tensor product of two vector representations, Littlewood-Richardson rule.*

5.17 Tensor product with the vector representation

As shown by G. Lusztig and M. Rosso [53,64] the representation theory of $U_q(\mathfrak{g})$ for a simple Lie algebra $\mathfrak{g}$ for generic values of q is the same as in the classical case $q = 1$. The tensor product of finite-dimensional representations of $sl(n + 1)$ is governed by the Littlewood-Richardson rule, cf. [55], which is in turn a consequence of the theory of characters and symmetric functions. The rule is essentially simple for the tensor product with vector representation, namely, we should add one box to the Young diagram so that in the result we obtain again a Young diagram, and if the column with $n+1$ boxes appears, it should be removed (as corresponding to trivial representation). The tensor product with vector representation is multiplicity free. For example, the tensor product of two vector representations of $sl(n + 1)$ decomposes as

$$L(\omega_1) \otimes L(\omega_1) = L(2\omega_1) \oplus L(\omega_2)$$

where $L(2\omega_1)$, $L(\omega_2)$ denotes the analog of the symmetric tensor, antisymmetric tensor, correspondingly, see Figure 9.

Let $P_{2\omega_1}, P_{\omega_2}$ denote the corresponding projectors. The PR-matrix is given as follows [40,41,18]:

$$PR = q^{-\frac{1}{2(n+1)}}(q^{\frac{1}{2}}P_{2\omega_1} - q^{\frac{-1}{2}}P_{\omega_2}).$$

So one can see that on antisymmetric tensors it acts as $-q^{-\frac{n+2}{2(n+1)}}$. Multiplying by $q^{\frac{n+2}{2(n+1)}}$ as in the previous section we get the coefficient of P_{ω_2} is -1.

Now consider the product of the finite-dimensional representation of weight μ with vector representation $L(\mu) \otimes L(\omega_1)$. And assume that we add the box to the s-th row of a Young diagram corresponding to μ. In particular, we assume that this is a correct operation. This means that we have variables of integration $\{t_i|i = 1,\ldots,s-1\}$ and the integral:

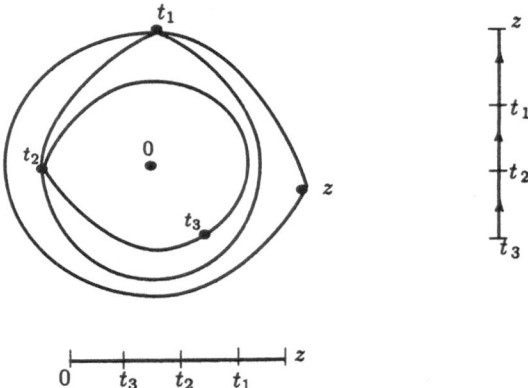

Figure 10: *Cycle for integration for vertex operator associated with the vector representation of $sl(n+1)$: contour for t_i starts and ends at $t_{i-1}, i = 2, 3, \ldots$; contour for t_1 starts and ends at z_1. In particular, all 'internal' contours are movable.*

$$\int \prod_{i=1}^{s-1} t_i^{\frac{(\mu, -\alpha_i)}{\varkappa}} \prod_{j<i}(t_i - t_j)^{\frac{(-\alpha_i, -\alpha_j)}{\varkappa}} \prod_{i=1}^{s-1}(z - t_i)^{\frac{(\omega_1, -\alpha_i)}{\varkappa}} \frac{dt_1}{t_1} \frac{dt_2}{t_2} \cdots \frac{dt_{s-1}}{t_{s-1}}$$

The natural domain of integration for asymptotic solution is

$$0 \leq t_{s-1} \leq t_{s-2} \leq \ldots \leq t_1 \leq z.$$

The leading asymptotic is equal to

$$z^{\frac{(\mu, -\sum_{i=1}^{s-1} \alpha_i) - (s-1)}{\varkappa}}.$$

The integral is easily taken using the formula for Euler's beta function:

$$\int_0^1 t^{\alpha-1}(1-t)^{\beta-1} dt = \frac{\Gamma(\alpha)\Gamma(\beta)}{\Gamma(\alpha+\beta)},$$

and the leading asymptotic coefficient is equal to

$$\prod_{p=1}^{s-1} \frac{\Gamma\left(\frac{(\mu, -\sum_{i=s-p}^{s-1} \alpha_i)}{\varkappa} - \frac{(p-1)}{\varkappa}\right) \Gamma(1 - \frac{1}{\varkappa})}{\Gamma\left(\frac{(\mu, -\sum_{i=s-p}^{s-1} \alpha_i)}{\varkappa} + 1 - \frac{p}{\varkappa}\right)}.$$

The above cycle for integration can be represented as the following vector of the tensor product of two dual Verma modules over the quantum group

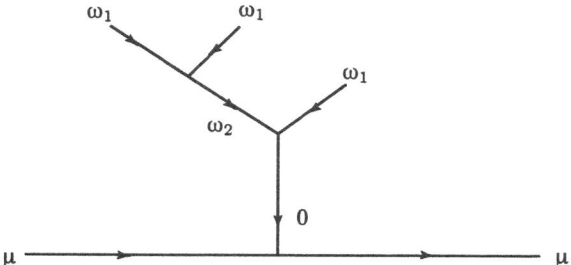

Figure 11: *Conformal block for zonal spherical function (A_2 case).*

$U_q(sl(n+1))$ (without Serre's relations) with highest weights μ and ω_1 as follows:

$$(-1)^{s-1}q^{\frac{-1}{4}(s-2)+\frac{1}{4}(\mu,\sum_{i=1}^{s-1}-\alpha_i)}(f_{s-1}\ldots f_1 v_0)^* \otimes v_1^* +$$

$$q^{\frac{1}{4}(s-1)}v_0^* \otimes (f_{s-1}\cdots f_1 v_1)^* +$$

$$\sum_{p=1}^{s-2}\left((-1)^{s-p-1}q^{\frac{-1}{4}(s-p-2)+\frac{1}{4}p+\frac{1}{4}(\mu,\sum_{p+1}^{s-1}-\alpha_i)}(f_{s-1}f_{s-2}\cdots f_{p+1}v_0)^* \right. \qquad (8)$$

$$\left. \otimes(f_p f_{p-1}\cdots f_1 v_1)^*\right)$$

The same formula holds if parameter μ is assumed to be generic (nonintegral). The above element of the tensor product of dual Verma modules is the image of the singular (vacuum) vector of the tensor product of irreducibles $L(\mu) \otimes L(\omega_1)$ of weight μ under the contravariant (Shapovalov) form S.

For generic μ and q, the tensor product $L(\mu) \otimes L(\omega_1)$ decomposes as

$$L(\mu) \otimes L(\omega_1) = L(\mu + h_1) \oplus L(\mu + h_2) \oplus \cdots \oplus L(\mu + h_{n+1}).$$

With the help of comultiplication Δ this might be iterated (in principle) and cycles for asymptotic solutions from ref. [44] can be represented as singular (vacuum) vectors of the tensor product of irreducible highest weight modules over the quantum group

$$L(\mu) \otimes L(\omega_1) \otimes \cdots \otimes L(\omega_1).$$

The space of singular (vacuum) vectors of weight μ of the tensor product

$$L(\mu) \otimes L(\omega_1) \otimes \cdots \otimes L(\omega_1)$$

with $n+1$ factors $L(\omega_1)$ is $(n+1)!$-dimensional, i.e., the situation here is the same as in A. Matsuo [15].

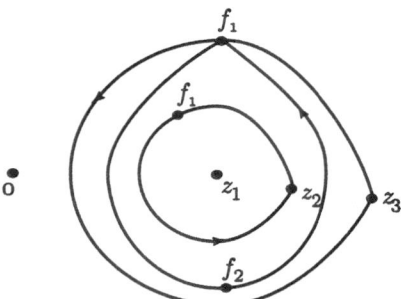

Figure 12: *Another contour for integration yielding the zonal spherical function in the case of the root system of type A_2.*

Remark 5.18   The fact that the cycle Δ corresponds to q-antisymmetric tensors is actually clear. Consider the tensor product of $n + 1$ vector representations of $sl(n + 1)$. Then the space of singular (vacuum) vectors of weight zero is one-dimensional. So there is only one way to get the variables of integration as needed: namely, to *add box under box, so that the Young diagram will be the column of $n+1$ boxes*, which corresponds to antisymmetric tensors, see Figure 11. One could also find different contours for integration giving the same homological class in $H_{*!, \frac{n(n+1)}{2}}(U, S^*)^-$ as cycle Δ (Figure 12), but cycle Δ is especially convenient for obtaining the Harish-Chandra expansion for the zonal spherical function of type A_n, and it has a geometric origin in elliptic coordinates.

Concluding remarks

The distinguished cycle Δ serves as a contour for integration for zonal spherical function of type A_n. It goes back to the classical calculation of I. Gelfand and M. Naimark of the zonal spherical function for $SL(n+1, \mathbb{C})$, originates in the so-called elliptic coordinates and provides a materialization of the flag manifold. The zonal spherical function corresponds to a particular conformal block of the WA_n-algebras. For more details about our approach from the point of view of conformal blocks of the WA_n-algebras see our ref. [46].

Acknowledgments. We are grateful to I. Gelfand for the suggestion to use the technique of ref. [27] in application to the Heckman-Opdam hypergeometric functions, to S. Lukyanov for discussions devoted to conformal field theory and W-algebras, to V. Schechtman for discussions concerning refs. [68,66,65], to V. Brazhnikov, M. Braverman for the discussions.

References

[1] Alexeev A., Faddeev L., Shatashvili S., Quantisation of symplectic orbits of compact Lie groups by means of functional integral, *Jour. of Geom. and Phys.*, **5** (1989), 391–406

[2] Alexeev A., Shatashvili S., From geometric quantization to conformal field theory, *Commun. Math. Phys.* (1990), 197–212

[3] Alvarez-Gaume L., Gomez C., Sierra G., Quantum group interpretation of some conformal field theories, *Phys. Lett. B* (1989), 142–151

[4] Aomoto K., Sur les transformation d'horisphère et les equations intégrales qui s'y rattachent, *J. Fac. Sci. Univ. Tokyo*, **14** (1967), 1–23

[5] Arnold V., The cohomology ring of the colored braid group, *Mat. Zametki*, **5** (1969), 227–23

[6] Awata H., Matsuo Y., Odake S., Shiraishi J., Excited states of Calogero-Sutherland Model and singular vectors of the W_n algebra, *Nucl. Phys.* **B449** (1995), p. 347

[7] Balog J., Feher L., O'Raifertaigh, Forgacz P., Wipf A., Toda theory and W-algebra from a gauged WZNW point of view, *Annals of Phys.*, **203** (1990), 76–136

[8] Belavin A., KdV equations and W-algebras, In: *Integrable Systems in Quantum Field Theory and Statistical Mechanics, Adv. Studies in Pure Math* **19** (1989)

[9] Berezin F., Laplace operators on semisimple Lie groups, *Trudy Mosk. Mat. ob-va*, **6** (1957), 371–463

[10] Berezin F., Gelfand I., Some remarks on the theory of spherical functions on symmetric Riemannian manifolds, *Tr. Mosk. Mat. O-va*, **5** (1956), 311–351

[11] Bernstein I. N., Gelfand I. M., Gelfand S. I., Structure of representations generated by vectors of highest weight, *Funct. Anal. and Appl.*, **5** (1971), 1–8

[12] Bilal A. Fusion and braiding in W-algebra extended conformal field theories (II): Generalization to chiral screened vertex operators labelled by arbitrary Young tableaux, *Intern. jour. of Modern phys. A*, **5**:10 (1990), 1881–1909

[13] Bouwknegt P., McCarthy J., Pilch K., Quantum group structure in the Fock space resolutions of $SL(n)$ representations, *Comm. Math. Phys.*, **131**, 125–156

[14] Cherednik I., Monodromy representations of generalized Knizhnik-Zamolodchikov equations and Hecke algebras, *Publ. RIMS Kyoto Univ.*, **27** (1991), 711–726

[15] Cherednik I., Integral solutions of trigonometric Knizhnik-Zamolodchikov equations and Kac-Moody algebras, *Publ. RIMS Kyoto Univ.*, **27** (1991), 727–744

[16] Cherednik I., A unification of Dunkl and Knizhnik-Zamolodchikov operators via affine Hecke algebras, *Invent. Math.*, **106** (1991), 411–431

[17] Dolotin V., On discriminants of multilinear forms, *Izvest. Math.*, **62**:2 (1998), 215–246

[18] Drinfeld V. G., Quantum groups, *Proc. ICM*, **1**, Berkeley, 1986, 798–820

[19] Fateev V. A., Lukyanov S. L., Poisson-Lie groups and classical W-algebras, *Intern. jour of Modern Physics A*, **7**:5 (1992), 853–876

[20] V. Fateev, S. Lukyanov., The models of two-dimensional conformal quantum field theory with Z_n symmetry, *Int. J. Mod. Phys A*, **3** (1988), 507–520

[21] Fateev V. A., Zamolodchikov A. B., Conformal quantum field theory models in two dimensions having $\mathbb{Z}_3$ symmetry, *Nucl. Phys.*, **B280** (1987), 644–660

[22] Felder G., BRST approach to minimal models, *Nucl. Phys.*, **B 317** (1989), 215–236

[23] Felder G., Wieczerkowski C., Topological representations of the quantum group $U_q(sl_2)$, *Comm. Math. Phys.*, **138** (1991), 583–605

[24] Feigin B., Frenkel E., Integrals of motion and quantum groups, *Springer Lecture Notes in Math.*, **1620** (1996), 349–418

[25] Finkelberg M., Schechtman V., *Localization of u-modules I. Intersection Cohomology of Real Arrangements*, hep-th 9411050

[26] Gelfand I., Spherical functions on symmetric Riemannian spaces, *Dokl. Akad. Nauk SSSR*, **70** (1950), 5–8

[27] Gelfand I. M., Naimark M. A., Unitary representations of classical groups, *Tr. Mat. Inst. Steklova*, **36** (1950), 1–288

[28] Gelfand I. M., Tsetlin M. L., Finite-dimensional representations of the group of unimodular matrices, *Dokl. Akad. Nauk SSSR*, **71** (1950), 825–828

[29] Gelfand I., Naimark M., Normed rings with involutions and their representations, *Izv. Akad. Nauk SSSR*, **12** (1948), 445–480

[30] Gelfand I., Raikov D., Irreducible unitary representations of locally bicompact groups, *Mat. sb.*, **13**:55 (1942), 301–316

[31] Gelfand I., Center of infinitesimal group ring, *Mat. Sb. Nov. Ser.*, **26**:28 (1950), 103–112

[32] Gelfand I., Dikii L., Fractional powers of operators and Hamiltonian systems, *Funct. Anal. and Appl.*, **10** (1976), 259–273

[33] Gindikin S. G., Karpelevich F. I., Plancherel measure for Riemannian symmetric spaces of nonpositive curvature, *Dokl. Akad. Nauk SSSR*, **145**:2 (1962), 252–255

[34] Gomez C., Sierra G., Quantum group meaning of the Coulomb gas, *Phys. Lett. B*, **240** (1990), 149–157

[35] Guillemin V., Sternberg S., The Gelfand-Zetlin system and quantization of the complex flag manifold, *Jour of Funct. Analysis*, **52** (1983), 106–128

[36] Harish-Chandra, Spherical functions on a semisimple Lie group I, *Amer. J. of Math*, **80** (1958), 241–310

[37] Heckman G., Opdam E., Root systems and hypergeometric functions I, *Comp. Math.*, **64** (1987), 329–352

[38] Heckman G., Hecke algebras and hypergeometric functions, *Invent. Math.*, **100** (1990), 403–417

[39] Helgason S., *Groups and Geometric Analysis*, Academic Press, Inc., 1984

[40] Jimbo M., Introduction to the Yang-Baxter equation, *Intern. Jour. of modern physics A*, **4**:15 (1989), 3759–3777

[41] Jimbo M., A q-analogue of $U(gl(N+1))$, Hecke algebra and Yang-Baxter equation, *Lett. in Math. Phys.*, **11** (1986)

[42] Kohno T., Quantized universal enveloping algebras and monodromy of braid groups, *Ann. Inst. Fourier (Grenoble)*, **37**:4 (1987), 139–160

[43] Kazarnovski-Krol A., Value of generalized hypergeometric function at unity, In: *Arnold-Gelfand Mathematical Seminars*, Birkhäuser Boston, pages 341–345, 1997

[44] Kazarnovski-Krol A., Cycles for asymptotic solutions and the Weyl group, In: *Gelfand Mathematical Seminars 1993–1995*, I. Gelfand, J. Lepowsky, M. Smirnov, eds., Birkhäuser Boston, pages 123–150, 1996

[45] Kazarnovski-Krol A., Harish-Chandra decomposition for zonal spherical functions of type A_n, In: *Arnold-Gelfand Mathematical Seminars*, Birkhäuser Boston, pages 347–359, (1997)

[46] Kazarnovski-Krol A., Matrix elements of vertex operators of the deformed WA_n-algebras and the Harish-Chandra solutions to Macdonald's difference equations, to appear in *Selecta Math.*, New series, **5** (1999), 1–45

[47] Kirillov A.N., Reshetikhin N., q-Weyl group and a Multiplicative Formula for Universal R-Matrices, *Commun. Math. Phys.*, **134** (1990), 421–431

[48] Koornwinder T., Orthogonal polynomials in two variables which are eigenfunctions of two algebraically independent partial differential operators 3, 4, *Indag. Math.*, **36** (1974), 357–381

[49] Kulish P. P., Reshetikhin N. Yu., Sklyanin E. K., Yang-Baxter equation and representation theory I, *Lett. Math. Phys.* **5** (1981), 393–403

[50] Kostant B., On the tensor product of a finite and infinite dimensional representation, *Jour of Funct. Anal.*, **20** (1975), 257–285

[51] Lukyanov S., Quantization of Gelfand-Dikii bracket, *Funct. Anal. and Appl.*, **22**:4 (1988), 1–10

52] Lukyanov S., Fateev V., Additional Symmetries and exactly soluble models in two-dimensional conformal field theory, *Sov. Sci. Rev. A Phys.*, **15** (1990), 1–117

53] Lusztig G., Quantum deformations of certain simple modules over enveloping algebras, *Adv. Math.*, **70** (1988), 237–249

54] Macdonald I., Commuting differential operators and zonal spherical functions, *Springer Verlag Lecture Notes in Math.*, **1271** (1987), 189–200

55] Macdonald I., *Symmetric Functions and Hall Polynomials, Second Edition*, Clarendon Press, Oxford University Press, 1995

56] Matsuo A., Integrable connections related to zonal spherical functions, *Invent. Math.*, **110** (1992), 95–121

57] Moore G., Reshetikhin N., A comment on quantum group symmetry in conformal field theory, *Nucl. Phys.*, **B328** (1989), 557–574

58] Opdam E., An analogue of the Gauss summation formula for hypergeometric functions related to root systems, *Math. Zeitschr.*, **212** (1993), 313–336

59] Olshanetsky M., Perelomov A., Quantum systems related to root systems and radial parts of Laplace operators, *Functional Analysis and its Appl.*, **12**:2 (1978), 57–65

60] Olshanetsky M., Perelomov A., *Explicit Solutions of Classical Generalized Toda Models*, pages 261–269, 1979

61] Rosso M., An analogue of P.B.W. Theorem and the universal R-matrix for $U_h sl(N+1)$, *Comm. Math. Phys.*, **124** (1989), 307–318

62] Ramirez C., Ruegg H., Ruiz-Altaba M., The Contour picture of quantum groups: Conformal field theories, *Nucl. Phys. B*, **364** (1991), 195–233

63] Ramirez C., Ruegg H., Ruiz-Altaba M., Explicit quantum symmetries of WZNW theories, *Phys. Lett. B* (1990), 499–508

64] Rosso M., Finite-dimensional representations of the quantum analog of the enveloping algebra of a complex simple Lie algebra, *Commun. Math. Phys.*, **117** (1988), 581–593

[65] Schechtman V., Varchenko A., Hypergeometric solutions of Knizhnik-Zamolodchikov equations, *Lett. Math. Phys.*, **20** (1990), 279–283

[66] Schechtman V., Varchenko A., Hypergeometric solutions of Knizhnik-Zamolodchikov equations, *Letters in Math. Phys.*, **20** (1990), 279–283

[67] Schechtman V., *Quantum groups and perverse sheaves. An example*, Stony Brook, preprint, September 1992

[68] Schechtman V., Varchenko A., Quantum groups and homology of local systems, IAS, preprint 1990, In: *Algebraic Geometry and Analytic Geometry*, Satellite ICM-90 conference, Springer-Verlag, 182–197

[69] Sekiguchi J., Zonal spherical functions on some symmetric spaces, *Publ. RIMS. Kyoto Univ.*, **12** (1977), 455–459

[70] Schechtman V., Varchenko A., Arrangements of hyperplanes and Lie algebra homology, *Invent. Math*, **106** (1991), pp. 139

[71] Todorov I., Quantum groups as symmetries of Chiral conformal algebras, *Lecture Notes in Phys.*, **370** (1990), 231–277

[72] Tsuchia A., Kanie Y., Vertex operators in Conformal field theory on P^1 and monodromy representations of Braid group, *Adv. Studies in Pure Math*, **16** (1988), 297–372

[73] Varchenko A., Multidimensional hypergeometric functions and their appearance in conformal field theory, algebraic K-theory, Algebraic geometry, In: *Proc. of International Congress of Mathematicians, Vol. I, II*, Kyoto, 1990, 281–300

[74] Varchenko A., Asymptotic solutions to the Knizhnik-Zamolodchikov equation and crystal base, *Comm. Math. Phys.*, **171** (1995), 99–138

[75] Varchenko A., Multidimensional hypergeometric functions and representation theory of Lie algebras and quantum groups; Advanced series, *Math. Physics*, **21** (1995)

[76] Weyl H., Harmonics on homogeneous manifolds, *Ann. of Math.*, **35** (1934), 486–499

[77] Zamolodchikov A. B., Infinite additional symmetries in two-dimensional conformal quantum field theory, *Theor. Math. Phys.*, **65**:3 (1986), 1205–1213 ——

[78] Zelevinsky A., Geometry and combinatorics related to vector partition functions, *Topics in Algebra* **26** part II (1990), 501–510

Department of Mathematics
Rutgers University
Piscataway, NJ 08854
e-mail:akrol@math.rutgers.edu

AMS Subject Classifications: Primary 22E30; Secondary 32G34, 33C80

The Existence of Fiber Functors

Alexander L. Rosenberg

Introduction

This paper contains a short proof of Deligne's theorem on inner characterization of Tannakian categories. The proof was given at Kazhdan's lecture course on topological and conformal field theories at Harvard during the 1990/1991 academic year.

The paper is essentially self-contained. The first section provides preliminaries on linear algebra in monoidal categories, which a reader new to the subject might consider as a series of (mostly quite simple) exercises. The topic of the second section is splitting objects and epimorphisms over appropriate faithfully flat algebra extensions. Then follows the Deligne theorem. Note that the Deligne's original argument [D] is based on a fragment of algebraic geometry in monoidal categories. The argument presented here uses elementary linear algebra, first facts of ring and module theory, in monoidal categories. All the statements (except of the theorem itself) have well known prototypes in conventional linear algebra which makes the proof transparent. Modulo this difference, the argument goes along the same lines as the original one in [D].

I would like to thank David Kazhdan for making me participate actively in a part of his course and for useful discussions on the subject.

1 Preliminaries

1.1. Monoidal categories Let $\mathcal{C}^\sim = (\mathcal{C}, \otimes, \alpha, l, r, 1)$, where $\mathcal{C}$ is a category, $\otimes$ is a functor from $\mathcal{C} \times \mathcal{C}$ to $\mathcal{C}$, α is a functor isomorphism $\otimes \circ (Id_\mathcal{C} \times \otimes) \to \otimes \circ (\otimes \times Id_\mathcal{C})$ (*associativity constraint*), and

$$l : Id_\mathcal{C} \to 1 \otimes Id_\mathcal{C}, \quad r : Id_\mathcal{C} \to Id_\mathcal{C} \otimes 1$$

are functor isomorphisms.

A morphism from $C^\sim = (C, \otimes, \alpha, \beta, l, r, 1)$ to $C'^\sim = (C', \otimes', \alpha', \beta', l', r', 1')$, otherwise called a *monoidal functor*, is a triple (F, ϕ, ϕ_0), where F is a functor $C \to C'$, $\phi = (\phi_{X,Y})$ is a functorial isomorphism, $\phi_{X,Y} : F(X) \otimes' F(Y) \to F(X \otimes Y)$ and $\phi_0 : F1 \to 1'$ an isomorphism such that the diagrams

$$
\begin{array}{ccccc}
F(X) \otimes' (F(Y) \otimes' F(Z)) & \xrightarrow{id_{F(X)} \otimes' \phi_{Y,Z}} & F(X) \otimes' F(Y \otimes Z) & \xrightarrow{\phi_{X,Y \otimes Z}} & F(X \otimes (Y \otimes Z)) \\
\alpha' \downarrow & & & & \downarrow F\alpha \\
(F(X) \otimes' F(Y)) \otimes' F(Z) & \xrightarrow{\phi_{X,Y} \otimes' id_{F(Z)}} & F(X \otimes Y) \otimes' F(Z) & \xrightarrow{\phi_{X \otimes Y,Z}} & F((X \otimes Y) \otimes Z)
\end{array}
\tag{1}
$$

and

$$
\begin{array}{ccc}
F(1) \otimes' F(X) & \xrightarrow{\phi_{1,X}} & F(1 \otimes X) \\
\phi_0 \otimes' id \downarrow & & \downarrow F(l_X) \\
1' \otimes F(X) & \xrightarrow{l'_{F(X)}} & F(X)
\end{array}
\qquad
\begin{array}{ccc}
F(X \otimes 1) & \xleftarrow{\phi_{X,1}} & F(X) \otimes' F(1) \\
F(r_X) \downarrow & & \downarrow id \otimes' \phi_0 \\
F(X) & \xleftarrow{r'_{F(X)}} & F(X) \otimes' 1'
\end{array}
$$

are commutative. The composition of morphisms is defined in an obvious way.

The tuple $C^\sim = (C, \otimes, \alpha, l, r, 1)$ is called a *monoidal category* if the triple $(L^\otimes : X \to X \otimes -, \alpha, l)$ is a morphism from $C^\sim$ to $(EndA, \circ, id, Id_A, id, id)$.

1.2. Examples (1) The category Vec_k of finite dimensional k-vector spaces with $\otimes = \otimes_k$.

(2) The category of finite dimensional regular representations (over a field k) of an affine algebraic group.

(3) The category of vector super-spaces.

(4) For a given k-linear category C, the category End_kC of k-linear functors $C \to C$ with the composition of functors as $\otimes$ and the identical functor as the unit object: $\otimes = \circ$, $1 = Id_C$.

1.3. Symmetric monoidal categories   Let $C^\sim = (C, \otimes, 1, \alpha, l, r)$ be a monoidal category. Then $C^{\sim\sigma} := (C, \otimes \circ S, 1, \alpha^{-1}, r, l)$, where S is the transposition functor

$$
C \times C \to C \times C, \quad (X, Y) \mapsto (Y, X),
$$

is a monoidal category called the *opposite monoidal category to* $C^\sim$.

A *symmetry* of monoidal category $C^\sim = (C, \otimes, 1, \alpha, l, r)$ is a functor isomorphism $\beta : \otimes \to \otimes \circ S$ such that $\beta S \circ \beta = id_\otimes$ and (Id_C, β, id_1) is a norphism from $\mathcal{C}^\sim$ to $\mathcal{C}^{\sim\sigma}$. A monoidal category with a fixed symmetry s called *symmetric*.

A morphism from a symmetric category $\mathcal{C}^\sim = (\mathcal{C}, \otimes, 1, \alpha, l, r; \beta)$ to ι symmetric category $\mathcal{C}'^\sim = (\mathcal{C}', \otimes', 1', \alpha', l', r'; \beta')$ is a morphism $F^\sim = F, \phi, \phi_0)$ from $\mathcal{C}^\sim$ to $\mathcal{C}'^\sim$ compatible with the corresponding symmetries. Γhe latter means that the diagram

$$
\begin{array}{ccc}
F(X) \otimes' F(Y) & \xrightarrow{\phi_{X,Y}} & F(X \otimes Y) \\
\beta'_{F(X),F(Y)} \Big\downarrow & & \Big\downarrow F\beta_{X,Y} \\
F(Y) \otimes' F(X) & \xrightarrow{\phi_{Y,X}} & F(Y \otimes X)
\end{array}
$$

:ommutes for all $X, Y \in Ob\mathcal{C}$.

..4. Algebras, modules and bimodules in monoidal categories

ʾix a monoidal category $\mathcal{C}^\sim = (\mathcal{C}, \otimes, 1, \alpha, l, r)$. An *algebra* (or *monoid*) in ʾ$^\sim$ is a pair (R, μ) where $R \in Ob\mathcal{C}$ and μ is a morphism $R \otimes R \to R$ such hat $\mu \circ (\mu \otimes id_R) \circ \alpha_{R,R,R} = \mu \circ (id_R \otimes \mu)$. The *unit* of an algebra (R, μ) is a norphism $\eta : 1 \to R$ such that $\mu \circ \eta \otimes id_R \circ l_R = id_R = \mu \circ id_R \otimes \eta \circ r_R$. The ınit (if it exists) is unique. We assume that all algebras considered here ιre unital. Algebras in $\mathcal{C}^\sim$ form a category which we denote by $Alg\mathcal{C}^\sim$.

A left module over an algebra (R, μ) is a pair (M, m), where $M \in Ob\mathcal{C}$, n is a morphism $R \otimes M \to M$ such that $m \circ id_R \otimes m = m \circ \mu \otimes id_M \circ \alpha_{R,R,M}$ ιnd $m \circ \eta \otimes id_M = id_M$. Left modules over $R^\sim = (R, \mu)$ form a category ₹$^\sim - mod$.

Similarly, we define the category $mod - R^\sim$ of right $R^\sim$-modules which are just left modules in the opposite monoidal category). A riple (m, M, m'), where (m, M) and (M, m') are resp. left and right $R^\sim$- ıodules, is called an $R^\sim$-*bimodule* if $m \circ id_R \otimes m' = m' \circ m \otimes id_R \circ \alpha_{R,M,R}$.

Suppose the functor $X \mapsto X \otimes -$ is right exact for any $X \in Ob\mathcal{C}$. Then here is a well defined functor $\otimes_R : mod - R^\sim \times R^\sim - mod \to \mathcal{C}$ which ssigns to any pair of resp. right and left $R^\sim$-modules $(M, m), (\nu, N)$ the okernel of the pair of morphisms $id_M \otimes \nu, m \otimes \nu \circ \alpha_{M,R,N} : M \otimes (R \otimes N) \to \Lambda \otimes N$. The functor $\otimes_R$ induces a structure of a monoidal category on he category $R^\sim - bimod$ of $R^\sim$-bimodules.

Let β be a symmetry of the monoidal category $\mathcal{C}^\sim$. An algebra $R^\sim = R, \mu)$ is called β-*commutative* (or *commutative* if β is fixed) if $\mu \circ \beta_{R,R} = \mu$. ʾhe full subcategory of $Alg\mathcal{C}^\sim$ formed by β-commutative algebras will be enoted by $Alg_\beta\mathcal{C}^\sim$.

For any β-commutative algebra $R^\sim$, the map $(m, M) \mapsto (m, M, m \circ$ ʾ$_{M,R})$ defines a functor, Δ_β, identifying the category $R^\sim - mod$ of left

$R^\sim$-modules with a full subcategory of the category $R^\sim - bimod$ of $R^\sim$-bimodules.

Suppose the functor $X \mapsto X \otimes -$ is right exact for any $X \in ObC$. Then the functor Δ_β identifies $R^\sim - mod$ with a monoidal subcategory of $R^\sim - bimod$. And the symmetry β induces a symmetry on $R^\sim - mod$.

1.4.1. Tensor algebras For any object V of the monoidal category $C^\sim$, the tensor powers of V are defined by $V^{\otimes 0} := 1$, $V^{\otimes 1} := V$, $V^{\otimes n} = V \otimes V^{\otimes n-1}$ for $n \geq 2$. Suppose that there exists a direct sum $\oplus_{n \geq 0} V^{\otimes n}$ and $\otimes$ preserves countable direct sums. Then the *tensor algebra*, $\mathbf{T}(V) = (T(V), \mu_V)$ of the object V is defined as follows: $T(V) = \oplus_{n \geq 0} V^{\otimes n}$, and the multiplication μ_V is given by the isomorphisms $V^{\otimes n} \otimes V^{\otimes m} \to V^{\otimes n+m}$ induced by the associativity constraint. The map $V \mapsto \mathbf{T}(V)$ is extended to a functor, $\mathbf{T}$, from the category C to the category $AlgC^\sim$ of algebras in $C^\sim$. This functor is left adjoint to the functor $AlgC^\sim \to C$, $(R, \mu) \mapsto R$, forgetting the multiplication.

1.4.2. Symmetric algebras Fix a symmetry β in a monoidal category $C^\sim$. Suppose that the functor $X \otimes -$ is right exact for all $X \in ObC$. Then for any algebra (R, μ) in $C^\sim$, the cokernel R_β of the pair of morphisms (multiplications) $\mu, \mu \circ \beta_{R,R} : R \otimes R \to R$ has a uniquely defined algebra structure, μ_β. The correspondence $(R, \mu) \mapsto (R_\beta, \mu_\beta)$ extends in a unique way to a *β-abelinization functor* $Ab_\beta : AlgC^\sim \to Alg_\beta C^\sim$ from the category of associative algebras to that of β-commutative algebras. The functor Ab_β is left adjoint to the inclusion $Alg_\beta C^\sim \to AlgC^\sim$.

Let V be an object of C such that there exists a direct sum $\oplus_{n \geq 0} V^{\otimes n}$, hence the tensor algebra of V. We define the *β-symmetric algebra* $\mathbf{S}_\beta(V)$ *of an object* V as the β-abelinization of the tensor algebra $\mathbf{T}(V)$.

1.5. Fragments of linear algebra

1.5.1. Inner hom Let V, W be objects of the monoidal category $C^\sim$. The *inner hom* from V to W (if any) is an object, $\mathfrak{Hom}(V, W)$ representing the functor $C(V \otimes -, W) : C^{op} \to Sets$. If $\mathfrak{Hom}(V, W)$ exists, then there is a canonical morphism

$$ev_{V,W} : V \otimes \mathfrak{Hom}(V, W) \to W \tag{1}$$

which is the image of $id_{\mathfrak{Hom}(V,W)}$ under the functorial isomorphism

$$C(\mathfrak{Hom}(V, W), \mathfrak{Hom}(V, W)) \to C(V \otimes \mathfrak{Hom}(V, W), W)$$

Conversely, any morphism $V \otimes X \to W$ determines a morphism $X \to \mathfrak{Hom}(V, W)$. Thus the canonical isomorphism $r_V : V \otimes 1 \to V$ induces a functorial bijection $C(V, W) \to C(1, \mathfrak{Hom}(V, W))$. In particular it defines a morphism $i_V : 1 \to \mathfrak{Hom}(V, V)$ corresponding to the identical morphism id_V.

The inner hom from an object V to the unit object 1 is denoted by $V^\vee$ and is called the *dual object to V* . The canonical arrow $ev_{V,1} : V \otimes V^\vee \to 1$ is called the *evaluation morphism* and is denoted simply by ev_V.

1.5.2. Finite objects Let V, W be objects of C such that $\mathfrak{Hom}(V, W)$ and $V^\vee$ exist. The canonical morphism $ev_V \otimes id_W : V \otimes (V^\vee \otimes W) \to W$ determines a morphism $\phi_{V,W} : V^\vee \otimes W \to \mathfrak{Hom}(V, W)$.

An object V is called *finite* if the morphism $\phi_{V,V}$ is an isomorphism.

1.5.2.1. Finite objects and monoidal functors Another characterization of finite objects: $V \in ObC$ is finite iff there exists an object $V^\vee$ and morphisms $ev : V \otimes V^\vee \to 1$ and $\gamma : 1 \to V^\vee \otimes V$ such that the compositions of $V \xrightarrow{V \otimes \gamma} V \otimes V^\vee \otimes V \xrightarrow{ev \otimes V} V$ and of $V^\vee \xrightarrow{\gamma \otimes V^\vee} V^\vee \otimes V \otimes V^\vee \xrightarrow{V^\vee \otimes ev} V$ are identical morphisms. The notations here are suggestive: $V^\vee$ is a dual object to V. It follows from this description that any monoidal functor (cf. 1.1) sends finite objects into finite objects and a dual object to a given finite object V to a dual object of the image of V.

Another consequence of this description is the following assertion: an object V is finite iff the morphism $\phi_{V,W} : V^\vee \otimes W \to \mathfrak{Hom}(V, W)$ is an isomorphism for all $W \in C$.

1.5.3. Trace and dimension Let $C^\sim = (C, \otimes, a, 1, ...)$ be a symmetric monoidal category with a symmetry β. Let V be a finite object of $C^\sim$. The β-trace of V (or simply *trace* of V if the symmetry β is fixed) tr_β is the composition of the canonical isomorphisms $\mathfrak{Hom}(V, V) \to V^\vee \otimes V$ and $\beta_{V^\vee, V} : V^\vee \otimes V \to V \otimes V^\vee$ and the evaluation morphism $ev_V : V \otimes V^\vee \to 1$.

The *dimension* of a finite object V is the composition of the trace and the canonical morphism $j_V : 1 \to \mathfrak{Hom}(V, V)$: $dim(V) = dim_\beta(V) := tr_\beta \circ j_V$.

1.5.3.1. Morphisms of symmetric monoidal categories and dimension It follows from 1.5.2.1 that any morphisms of symmetric monoidal categories $(F, \phi, \phi_0) : (C, \otimes, a, 1, l, r, \beta) \to (C', \otimes', a', 1', l', r', \beta')$ preserves dimensions, i.e., for any finite object V of C, $dim_\beta(V) = dim_{\beta'}(F(V))$.

1.5.4. Symmetric powers   The *symmetric power*, $S_\beta^n(V)$, of an object V is the image of the symmetrization $s = \sum_{\sigma \in S_n} \beta_\sigma(V) : V^{\otimes n} \to V^{\otimes n}$, where $\beta_\sigma(V)$ denotes the automorphism of $V^{\otimes n}$ determined by the permutation σ and the symmetry β.

Suppose that $C^\sim$ is a k-category, where k is a field of zero characteristic. Then $S_\beta^n(V)$ is also the image of $s/n!$. And the dimension of $S_\beta^n(V)$ is given by the usual formula:

$$dim S_\beta^n(V) = tr(s/n!) = dim(V)(dim(V)+1)...(dim(V)+n-1)/n! \quad (1)$$

The symmetric algebra $S_\beta(V)$ (cf. 1.4.2) is naturally isomorphic to $(\oplus_{n \geq 0} S_\beta^n(V), \mu)$.

1.5.5. Exterior powers   The *exterior power*, $\wedge^n(V)$ of an object V is the image of the antisymmetrization $a = \sum_{\sigma \in S_n} (-1)^{\epsilon(\sigma)} \beta_\sigma(V) : V^{\otimes n} \to V^{\otimes n}$, where $\beta_\sigma(V)$ is the same as in 1.5.4. In the case of zero characteristic, $\wedge^n(V)$ is also the image of $a/n!$. In this case the dimension of $\wedge^n(V)$ is given by the conventional formula:

$$dim(\wedge^n(V)) = tr(a/n!) = dim(V)(dim(V)-1)...(dim(V)-n+1)/n! \quad (1)$$

2   Splitting objects and morphisms in a monoidal category

Fix a monoidal category $C^\sim = (C, \otimes, a, 1)$ and its symmetry β.

2.1. Lemma   *Let M be a finite object of $C^\sim$ such that $dim(M)$ is not a nonpositive integer. Then there exists a faithfully flat algebra B such that the B-module $B \otimes M$ admits a decomposition into a direct sum $B \oplus N$ of B-modules.*

Proof.   (a) Consider the functor G which assigns to any algebra $\mathcal{R} = (R, \mu)$ in $C^\sim$ the set of all pairs of R-module morphisms $R \xrightarrow{u} R \otimes M \xrightarrow{v} R$. The functor G is isomorphic to the functor G' which assigns to any algebra R the set of all pairs of morphisms $v' : M \to R$, $u' : M^\wedge \to R$. Since $C(M, R) \simeq Alg(S_\beta(M), R)$, the functor G' is isomorphic to

$$Alg(S_\beta(M \oplus M^\wedge), R) \simeq Alg(S_\beta(M) \otimes S_\beta(M^\wedge), R)$$

(b) The composition $(u, v) \mapsto v \circ u$ is a functor morphism

$$c(R) : G(R) \to R - mod(R, R) \simeq C(1, R) \simeq Alg(S_\beta(1), R)$$

Let c' denote the corresponding algebra morphism $S_\beta(1) \to S_\beta(M) \otimes S_\beta(M^\wedge)$.

(c) Let Φ be a subfunctor of the functor G which assigns to any algebra R the subset of $G(R)$ consisting of all pairs (u, v) such that $v \circ u = id_R$. The functor $R \mapsto id_R$ is corepresentable by the algebra 1. Hence Φ is corepresentable by the fiber product B of $1 \xleftarrow{\delta} S_\beta(1) \xrightarrow{c'} S_\beta(M) \otimes S_\beta(M^\wedge)$, where δ is the morphism identifying each homogeneous component of $S_\beta(1)$ with 1. The composition of c' with the embedding $i_1 : 1 \to S_\beta(1)$ identifying 1 with the first component of $S_\beta(1)$ is the composition of the canonical morphism $\gamma : 1 \to M \otimes M^\wedge$ and the embedding $M \otimes M^\wedge \to S_\beta(M) \otimes S_\beta(M^\wedge)$. Therefore B is the biggest among quotients of $S_\beta(M) \otimes S_\beta(M^\wedge)$ which coequalize γ and $1 \to S_\beta(M) \otimes S_\beta(M^\wedge)$. It follows that

$$B \simeq \oplus_{m \in \mathbb{Z}} S_\beta^n(M) \otimes S_\beta^{n+m}(M^\wedge),$$

where the transition arrows are multiplications by γ.

(d) The unit $1 \to B$ splits in the category $\mathcal{C}$.

In fact, since $char(k) = 0$, S_β^n is a direct summand of $\otimes^n$ determined by the projection $s/n!$, $s := \sum_{\sigma \in S_n} \beta_\sigma$. In particular, the pairing between $\otimes^n(M^\wedge)$ and $\otimes^n(M)$ induces a pairing (duality) $ev_{S,n} : S_\beta^n(M^\wedge) \otimes S_\beta^n(M) \to 1$. Set $\tau_0 = id$ and $\tau_n = ev_n/d_n$, where $d_n := d(d+1)...(d+n-1)/n!$ is the dimension of $S_\beta^n(M)$. The morphisms τ_n commute with multiplication by γ, $\gamma_n : S_\beta^n(M^\wedge) \otimes S_\beta^n(M) \to S_\beta^{n+1}(M^\wedge) \otimes S_\beta^{n+1}(M)$, $\tau_{n+1} \circ \gamma_n = \tau_n$. Hence morphisms $\{\tau_n | n \geq 0\}$ define a morphism τ from the direct summand (and a subalgebra) $B' = colim(S_\beta^n(M^\wedge) \otimes S_\beta^n(M))$ of the algebra B to 1. The morphism τ is left inverse to the unit $1 \to B$. $\square$

2.2. Lemma *Let $v : M \to N$ be an epimorphism of finite objects. Then there exists a faithfully flat algebra B such that the morphism $id_B \otimes v : B \otimes M \to B \otimes N$ splits.*

Proof. (a) Suppose that $N = 1$, and let $v^\wedge$ denote the dual to v morphism $1 \to M^\wedge$. Let B denote the fiber product of the pair of arrows $1 \xleftarrow{} S_\beta(1) \xrightarrow{Sv^\wedge} S_\beta(M^\wedge)$. Clearly $B \simeq colim S_\beta^n(M^\wedge)$, where translation arrows $S_\beta^n(M^\wedge) \to S_\beta^{n+1}(M^\wedge)$ are the multiplication by $v^\wedge$. The canonical morphism $\gamma : 1 \to M^\wedge \otimes M$ defines a B-module morphism $\phi : B = B \otimes 1 \to B \otimes B \otimes M$ which is right inverse to $id_B \otimes v$.

Note that a monomorphism $V \to W$ of finite objects induces a filtration on $\otimes^n(W)$. Since the restriction of the functor $\otimes^n$ to the subcategory of finite modules is exact, the associated graded module is $\otimes^n(V \otimes W/V)$. Applying the projection $s = (\sum_{\sigma \in S_n} \sigma)/n! : \otimes^n \to S_\beta^n$, we obtain the corresponding filtration of $S_\beta^n(W)$ with the quotients $S_\beta^i(V) \otimes S_\beta^{n-i}(W/V)$, $0 \leq i \leq n$. Applying this to the monomorphism $v^\wedge : 1 \to M^\wedge$, we obtain the filtration

$$0 \to 1 = S_\beta^0(M^\wedge) \xrightarrow{v^\wedge} M^\wedge = S_\beta^1(M^\wedge) \xrightarrow{v^\wedge} ... \xrightarrow{v^\wedge} S_\beta^n((M^\wedge)$$

with quotients $S_\beta^i(M^\wedge/1)$. This shows that the algebra B is faithfully flat.

(b) Let now $v : M \to N$ be an arbitrary epimorphism of finite objects. Since N is finite, in particular the functor $\mathfrak{Hom}(-, N)$ is exact, the morphism $v' = \mathfrak{Hom}(v, id_N)\mathfrak{Hom}(M, N) \to \mathfrak{Hom}(N, N)$ is an epimorphism. Let M' be the fiber product of the morphism v' and the canonical morphism $1 \to \mathfrak{Hom}(N, N)$. Since the category C is abelian, the projection $\pi : M' \to 1$ is an epimorphism. Given a commutative algebra B, the morphism $id_B \otimes v$ splits iff the morphism $id_B \otimes \pi$ splits. Thus the assertion follows from the part (a) of the argument.   □

3   Deligne's theorem

Given a k-linear symmetric monoidal category $C^\sim$, a fiber functor over a commutative k-algebra R is a k-linear morphism (F, ϕ, ϕ_0) from $C^\sim$ to the symmetric monoidal category of R-modules over a commutative k-algebra R such that the functor F is exact and faithful. The monoidal category $C^\sim$ is called *Tannakian* if it admits a fiber functor over some nonzero commutative ring.

3.1.   Theorem   *Let $C^\sim = (C, \otimes, 1, \beta)$ be a rigid symmetric monoidal category over a field k of zero characteristic such that $C(1, 1) = k$. Then the following conditions are equivalent:*

(a) The monoidal category $C^\sim$ is Tannakian.

(b) The dimension of each object of $C^\sim$ is a nonnegative integer.

(b') The dimension of each nonzero object of $C^\sim$ is a positive integer.

(c) For each object X of C, there exists an integer n such that $\wedge^n(X) = 0$.

Proof.   (a) $\Rightarrow$ (b). By 1.5.2.1 and 1.5.3.1 fiber functors send finite objects to finite objects and preserve dimensions. Finite objects in the category of vector spaces are finite dimensional vector spaces and the dimension in the sense of 1.5.3 coincides in Vec_K with the usual dimension over K, hence the assertion.

$(b) \Leftrightarrow (c)$. This follows from the formula 1.5.5(1).

$(b) \Rightarrow (b')$. Suppose that V is a nonzero object of $\mathcal{C}$ such that $dim(V) = 0$. Since $V \neq 0$, the canonical morphism $\eta_V : 1 \to V^\vee \otimes V$ is nonzero, hence injective. Let W be a cokernel of η_V. We have: $dim(W) = dim(V^\vee \otimes V) - dim(1) = dim(V^\vee)dim(V) - 1 = -1$ which contradicts the assumption that $dim(W) \geq 0$.

The implication $(b') \Rightarrow (b)$ is obvious.

$(b')\&(c) \Rightarrow (a)$ (i) Any rigid abelian (symmetric) monoidal category $\mathcal{C}^\sim$ is canonically embedded into an abelian monoidal symmetric category $\mathcal{C}'^\sim = (\mathcal{C}', \otimes', a', 1, ...)$ such that $\mathcal{C}'$ has colimits and $\mathcal{C}$ is the full subcategory of $\mathcal{C}'$ generated by all finite objects of $\mathcal{C}'^\sim$. Namely the category $\mathcal{C}'$ is the category $Ind\mathcal{C}$ of *ind-objects* of $\mathcal{C}$ (SGA4, I.8.2). Recall that objects of $Ind\mathcal{C}$ are filtered inductive systems (V_i) in $\mathcal{C}$ and morphisms from (V_i) to (W_j) are given by $\mathcal{C}'((V_i), (W_j)) = lim_i colim_j \mathcal{C}(V_i, W_j)$. The bifunctor $\otimes$ induces a bifunctor $\otimes'$ which is also exact. The associativity constraint a and the symmetry constraint β induce resp. an associativity and symmetry constraints in $\mathcal{C}'^\sim$.

(ii) Fix a finite object X of the monoidal category $\mathcal{C}^\sim$. And let $dim(X) = d$ be an positive integer. By 2.1, there exists a faithfully flat algebra R in $\mathcal{C}'^\sim$ such that $R \otimes X \simeq (d)X \oplus N$ for some R-module N. For any two objects, V and W, we have: $\Lambda^n(V \oplus W) \simeq \oplus_{i+j=n} \Lambda^i(V) \otimes \Lambda^j(W)$. In particular, N is a direct summand in $\Lambda^{d+1}(R \otimes X) = R \otimes \Lambda^{d+1}(X) = 0$, hence $N = 0$.

(iii) It follows from (i) that there exists a commutative faithfully flat algebra B in $\mathcal{C}'^\sim$ such that $B \otimes X \simeq (dimX)B$ for any finite object X of $\mathcal{C}^\sim$. By 2.2, there exists a faithfully flat B-algebra R in $\mathcal{C}'^\sim$ such that for any short exact sequence $\mathcal{E}$ in $B - mod$, the sequence $R \otimes_B \mathcal{E}$ splits. It follows that the functor $X \mapsto \mathcal{C}'(1, R \otimes X)$ is a fiber functor over the commutative k-algebra $\mathcal{C}'(1, R)$. $\square$

References

[D]   P. Deligne, Catégories Tannakiennes, *Grothendieck Festschrift*, v. II, Birkhäuser, 1990.

[DM] P. Deligne and J. S. Milne, Tannakien Categories in *Hodge Cycles, Motives, and Shimura Varieties*, LNM 900, Springer Verlag, 1982, pp. 101–228.

[M]   S. MacLane, *Categories for the Working Mathematicians*, Graduate Texts in Mathematics 5, Springer Verlag, 1971.

[S]   S. Saavedra, *Catégories Tannakiennes*, LNM 265, Springer-Verlag, 1972.

Department of Mathematics
Kansas State University
137 Cardwell Hall
Manhattan, Kansas 66506
e-mail:rosenber@math.ksu.edu

AMS Subject Classification: 17Bxx